The Way
Life
Works

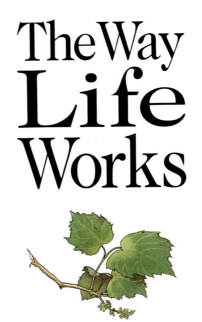

The Way
Life
Works

Mahlon Hoagland · Bert Dodson

TIMES 𝕿 BOOKS

RANDOM HOUSE

To

Bernard D. Davis

and

Judith Church

Library of Congress Cataloging-in-Publication Data

Hoagland, Mahlon B.
        The way life works / Mahlon Hoagland and Bert Dodson. — 1st ed.
                p.   cm.
        Includes index.
        ISBN 0-8129-2020-1
        1. Life (Biology)   I. Dodson, Bert.   II. Title.
        QH501.H57   1995
        574 — dc20

Manufactured in the United States of America
9 8 7 6 5 4 3 2
First Edition

Book Design: The Laughing Bear Associates, Montpelier, Vermont

AUTHORS' NOTE

When we — biologist and artist — first met in 1988, we discovered that we shared a fascination with the unity of life — how, deep down, all living creatures, from bacteria to humans, use the same materials and ways of doing things.

We began exploring ways we might share our wonder with others, and came to believe we could achieve our purpose through an intimate merging of science and art. In the process, we hoped to persuade our audience that a deeper understanding of nature would enhance their appreciation of its beauty — and thereby enrich their lives.

Scientist as teacher and artist as student explained, questioned, searched, and argued. One day, Bert emerged with a two-page spread of pictures and Mahlon got a new vision of what he thought he knew; artist became teacher, scientist became student. Our confidence grew. We sifted, sorted, and pieced together our interpretation of the way life works.

The scientist wants to leave the reader with a feeling of awe and pride in the achievements of scientific exploration, in the human potential for ever deeper understanding. The artist, on the other hand, sees the possibility that an appreciation for our oneness with the living world can guide our individual actions as we shape our collective future. We hope our readers will be moved by both.

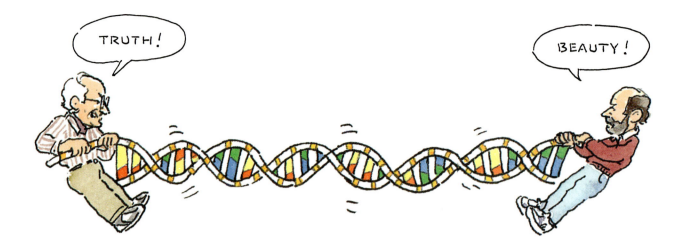

ACKNOWLEDGMENTS

We are grateful to these colleagues and friends for criticizing parts of the book and educating us in the process: Nancy Bucher, William Crane, Liz Davis, Jerry Gross, Bill Layton, Beth Luna, Ernst Mayr, Javier Penalosa, Sheldon Penman, Oscar Scornik, Walter Stockmeyer, Kip Sluder, Bernie Trumpower, and George Witman.

Special thanks are due Thoru Pederson, Judy Hauck, and Scott Dodson for their enduring interest, encouragement, and valued suggestions. Sue Ricker provided exceptional secretarial help over the four years the book took shape. Bonnie Dodson, John Stephens, and Moria Stephens gave valuable assistance with the art.

Betsy Rapoport, our editor at Times Books, gave us sound guidance and sharpened our sensitivity to the needs of the general reader, while Sam Vaughan, our editor at Random House, provided upbeat and cheering criticism and advice; and Jane Hoover of Lifland et al. Bookmakers, provided superb copy editing.

We are especially appreciative of The Laughing Bear Associates — Mason Singer, Rachel Goldenberg, Bob Nuner, and Linda Mirabile — for their patient and imaginative commitment to every detail of the book's design.

Jill Kneerim, our agent at Palmer & Dodge, regularly boosted our morale with her eloquently expressed enthusiasm. We're profoundly grateful for the help of the late Lewis Thomas in obtaining the financial support of the Richard Lounsbery Foundation, and pleased by the willingness of David Goudy, director of The Montshire Museum of Science, to have the museum act as sponsoring institution and to arrange exhibits of the book's paintings.

Finally, we feel deeply indebted to our wives, Tess Hoagland and Bonnie Dodson, for their loving support and practical assistance.

The capacity to tolerate complexity and welcome contradiction, not the need for simplicity and certainty, is the attribute of an explorer. Centuries ago, when some people suspended their search for absolute truth and began instead to ask how things worked, modern science was born. Curiously, it was by abandoning the search for absolute truth that science began to make progress, opening the material universe to human exploration. It was only by being provisional and open to change, even radical change, that scientific knowledge began to evolve. And ironically, its vulnerability to change is the source of its strength.

— Heinz R. Pagels in Perfect Symmetry:
The Search for the Beginning of Time

…And it is a strange thing that most of the feeling we call religious, most of the mystical outcrying which is one of the most prized and used and desired reactions of our species, is really the understanding and the attempt to say that man is related to the whole thing, related inextricably to all reality, known and unknowable. This is a simple thing to say, but a profound feeling of it made a Jesus, a St. Augustine, a Roger Bacon, a Charles Darwin, an Einstein. Each of them in his own tempo and with his own voice discovered and reaffirmed with astonishment the knowledge that all things are one thing and that one thing is all things — a plankton, a shimmering phosphorescence on the sea and the spinning planets and an expanding universe, all bound together by the elastic string of time.

— John Steinbeck in Log from the Sea of Cortez

CONTENTS

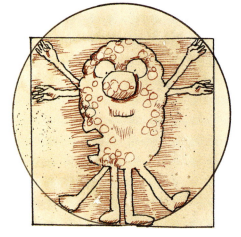

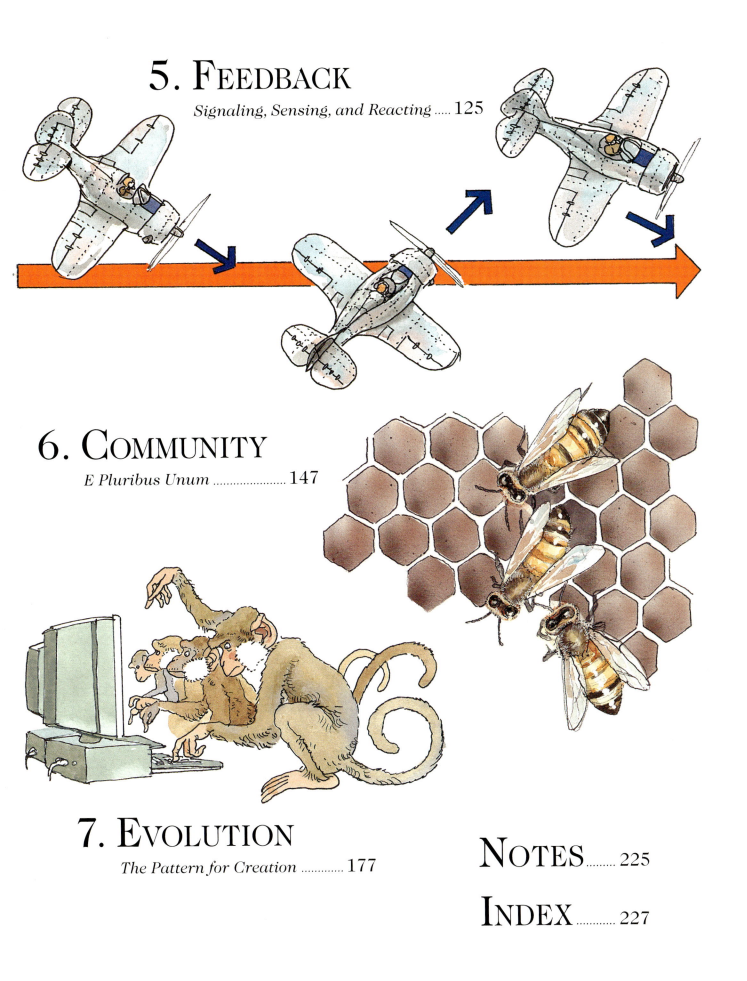

5. FEEDBACK
Signaling, Sensing, and Reacting 125

6. COMMUNITY
E Pluribus Unum 147

7. EVOLUTION
The Pattern for Creation 177

NOTES 225

INDEX 227

INTRODUCTION

Imagine you are walking along a deserted beach and you come upon the carcass of a whale. Time and tide and carrion birds have taken much of the flesh. Your first reaction might be a compassionate recognition of kinship. You might be curious about what happened — what was this whale's story?

As you examine the skeleton, a pattern strikes you. In each of the whale's front fins, the bones are arranged in three sections, with one bone in the section closest to the body, two parallel bones in the middle section, and five radiating branches of smaller bones in a more complex outer section. In fact, the bones of a whale's fin look very much like those of a human arm and hand. The proportions differ, but the pattern is remarkably similar.

How is it that a whale has arms like yours? And why does a whale have finger bones when it doesn't have fingers? Does this mean we're related to whales? Could it be that this limb pattern has been around longer than either whales…or humans?

A SINGULAR THEME

When we muse about life, what impresses us is its diversity — the sheer variety of organisms everywhere we look. Television programs and books about nature tend to celebrate the astonishing multiplicity of ways that life has adapted to our planet. This book's theme is different: It celebrates unity. It focuses on the things common to all forms of life, everywhere on earth.

Those homologous, or common, patterns in the bones of the human arm and the whale fin and, for that matter, in the bones of a bird's wing and a bat's wing, and even in the fossil remains of creatures that lived millions of years ago — are the first visible signs of unity. And the deeper we explore, the more signs we discover.

Every living being is either a cell or is made of cells: tiny, animate entities that gather fuel and building materials, produce usable energy, and grow and duplicate.

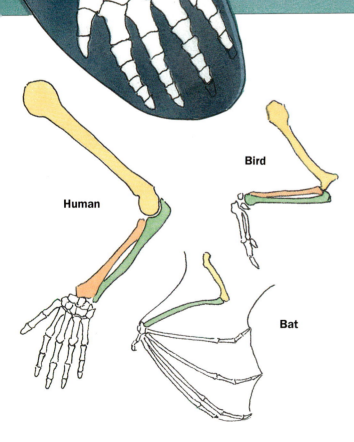

Human

Bird

Bat

And inside all living cells — cells as different as those of bacteria, or flies, or frogs, and humans, and those of skin, liver, and brain— are the same, or nearly the same, molecules and interactions, that make life work.

We are led to twin conclusions: The basic structures and mechanisms that sustain life on earth today are common to all living creatures; and the processes that have created life as we know it have been guided by a common set of rules.

Thus, all forms of life are connected to one another and to their predecessors — all the way back to what was most probably a single beginning, almost 4 billion years ago.

We believe that the beauty of living things becomes ever more breath-taking as you come to appreciate the patterns that bind them all together.

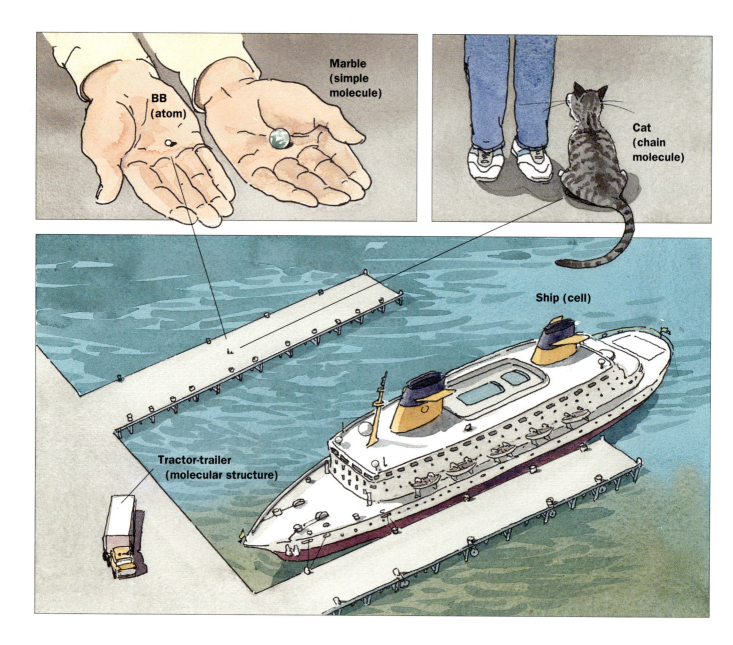

BB
(atom)

Marble
(simple
molecule)

Cat
(chain
molecule)

Ship (cell)

Tractor-trailer
(molecular structure)

THINKING SMALL

Much of this book takes place inside the cell. If you are unfamiliar with this microscopic landscape, understanding just how small and how numerous molecules are requires a considerable stretch of the imagination.

The great Scottish mathematician and physicist Lord Kelvin said:"Suppose that you could mark the molecules in a glass of water; then pour the contents of the glass into the ocean and stir the latter thoroughly so as to distribute the marked molecules uniformly throughout the seven seas; if then you took a glass of water anywhere out of the ocean, you would find in it about a hundred of your marked molecules."

Size and speed are related. Generally, the smaller an object is, the faster it can move. Water molecules, and all the other thousand or so kinds of molecules you have within you, swim about at stupendous speeds, flashing past each other and bumping into each other every millionth of a millionth of a second.

Life depends upon these frequent and vigorous collisions. It becomes a little easier to grasp the speed of the life-sustaining chemical transformations that constantly occur inside your cells (at the rate of thousands of events per second) when you realize that the participants move and collide millions of times faster.

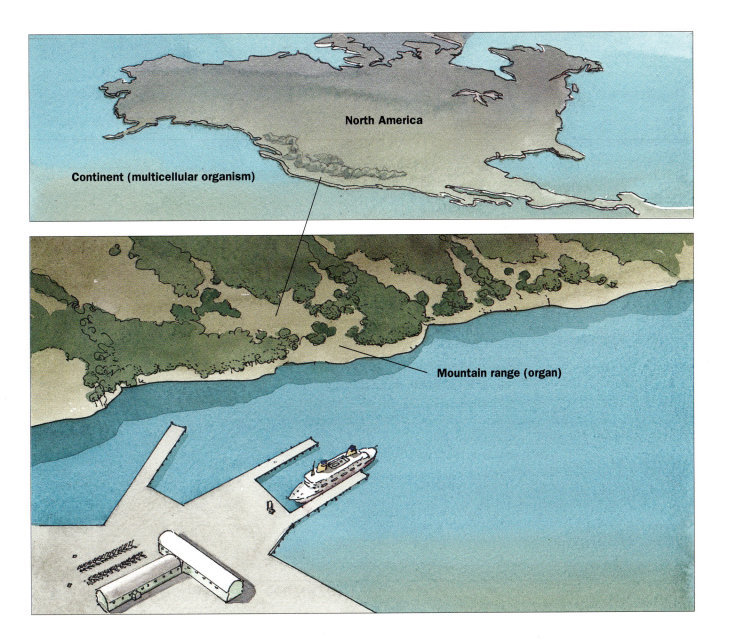

Continent (multicellular organism)

North America

Mountain range (organ)

When we think about parts of the body, we tend to think of muscles, heart, brain, etc. The next step down in size brings us to the cells of which those parts are made. That drop in size is immense. Human cells are about ten times smaller than the point of a pin, and your body is composed of 5 trillion of them. And within each cell are multitudes of atoms, molecules, and structures made of molecules — the principle characters in our story.

As we introduce them, the picture above might help you to grasp their relative sizes.

Imagine you are standing on a pier. In one hand, you hold a BB — its size will represent an atom. In the other hand, you hold a marble — analogous to a simple molecule. Next to you is a cat — a chain molecule. Parked nearby is a tractor-trailer truck — a molecular structure. And tied up at the pier is an ocean liner — a cell. The pier is on the coast of North America — the whole continent being analogous in size to a human being.

On the following four pages we present a visual guide for distinguishing small things. Notice that four separate scales are necessary for spanning the range of size from atom to cell (a 200,000 fold jump).

From Atoms to Cells — Comparative Sizes

Scale 1. Atoms and Molecules
Magnified 50 Million Times

Hydrogen · **Carbon** · **Nitrogen** · **Oxygen** · **Phosphorus** · **Sulfur**

Atoms are the elemental units of which everything in the universe, living and non-living, is made. Atomic diameters range from one to a few hundred millionths of an inch.

Molecules are atoms bonded together. Much of life depends on three tiny molecules that have 2 to 3 atoms apiece: carbon dioxide (CO_2), the ultimate source of life's carbon atoms; oxygen (O_2), the gas crucial to energy generation in most life forms; and water (H_2O), the sea inside our cells in which life's machinery is bathed, and which aids chemical events inside our cells.

Carbon dioxide · **Water** · **Oxygen (gas)**

Roughly one thousand different kinds of slightly larger molecules made of 10 to 35 atoms are also found inside cells. These small molecules are either food (fuel) or building materials, or molecules that have been or will be food or building materials. We call all these *simple molecules*. The important ones in this book are nucleotides, amino acids, and sugars, whose colors and shapes are depicted here to reflect their functions.

Protein

Nucleotide

Amino acid

Sugar

SCALE 2. CHAIN MOLECULES
Magnified 10 Million Times

The vital working parts inside cells are chain molecules — very long strings of many simple molecules linked to one another. The most numerous of the chain molecules are proteins, which consist of 300 to 400 or more amino acids strung end to end. Each kind of protein molecule — there are thousands of different kinds — has a special job to do in the cell. Cells also contain many varieties of *ribonucleic acid* (RNA), which can have tens of thousands of linked nucleotides, and *deoxyribonucleic acid* (DNA), which can have of millions of nucleotides.

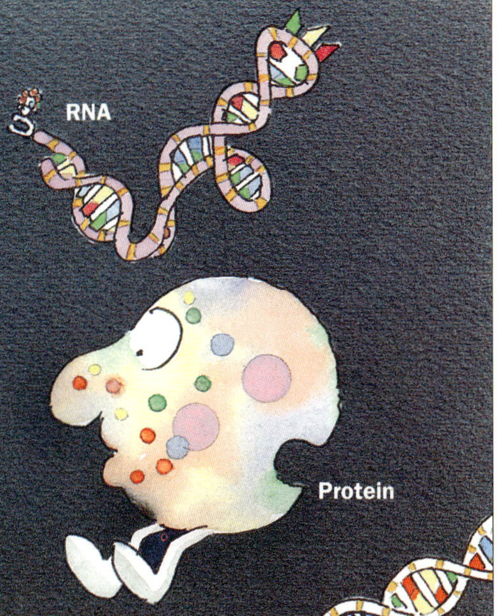

RNA

Protein

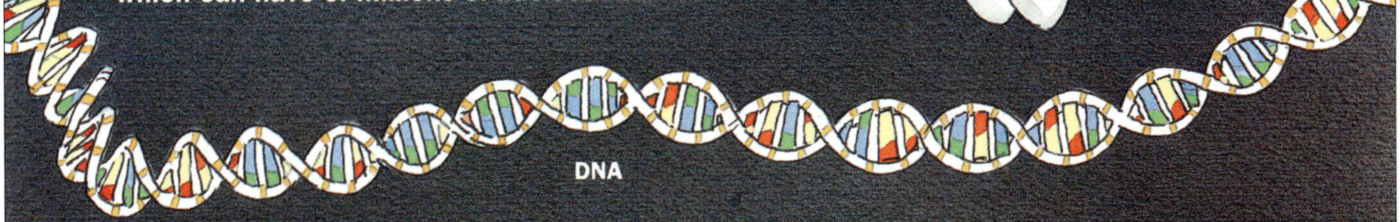

DNA

SCALE 3. MOLECULAR STRUCTURES
Magnified 1 Million Times

Chain molecules can fit together inside a cell in complex architectural arrangements called *molecular structures.* These are the cell's infrastructure, the equivalent of its roads, tunnels, power plants, factories, and libraries. Shown here are a ribosome, the cell's protein-making factory, and a bit of a mitochondrion, the cell's energy generator.

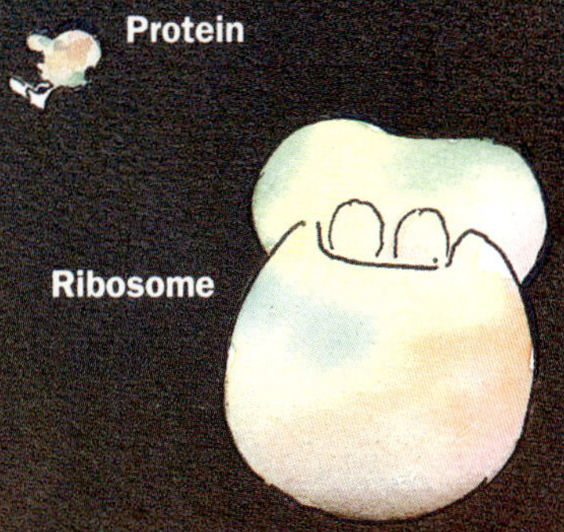

Protein

Ribosome

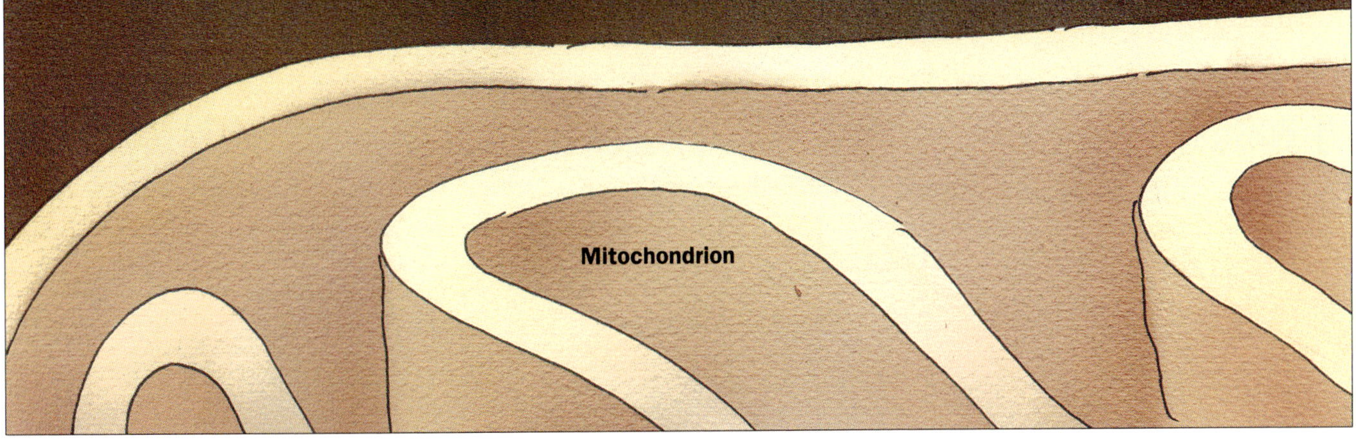

Mitochondrion

Bacterium

Animal cell

Mitochondrion

Ribosomes

Scale 4. A Cell
Magnified 10 Thousand Times

An animal cell, like this one, has a nucleus,
which contains most of its DNA. The nucleus is
surrounded by the cytoplasm, where most of the
cell's active processes occur. An average plant
cell is about three times larger than an animal cell.

PARTS AND WHOLES

It's useful to think of life's organization in levels, from the simple to the complex: atoms, simple molecules, chain molecules, molecular structures, cells — and onward and upward to organs, organisms, and communities of organisms. A higher level includes everything in the levels below it, as shown by the Russian dolls above.

Scientists find that knowing a lot about a lower level produces useful explanations of what's happening at the next higher level. To understand how your car works, you must know something about cylinders and spark plugs and fuel injection and how they interact.

This way of getting to understand the whole by learning about its parts, called reductionism, has produced in the last several decades an explosion of knowledge about what genes are and how they work, and how living processes are energized, informed, operated, and controlled. They are the "what" and "how" questions we take up in the first six chapters.

When we ask why things are the way they are we need to see things from the outside, and in relationship to others and to the surroundings. For example, why do birds have different beaks? To discover the answer, we need to study not just the birds themselves but, among other things, the food they eat. "Why" questions address patterns of connection in both space and time. They relate particularly to evolution — the subject of our final chapter.

Biochemists and molecular biologists tend to see themselves as reductionists, while naturalists and ecologists tend to take a holistic view. But, in fact, every scientist must shift his or her gaze regularly from the parts to the whole — from the trees to the forest — and back again.

We recommend that you try to be similarly fluid so that you can move back and forth with us as we shift from the tiny micro world to the larger macro world and back again.

Protein

DNA

THE WAY LIFE WORKS — THE BASIC IDEA

In exploring life's unity, we set out to connect the world of molecules with the world you can see around you.

Our central characters in this story are two chain molecules: One carries the information, the other does the work. To put things simply, you might say that life is played out in the inter-action between these two players — DNA and protein, whose relationship can be seen as that between information and machinery.

PICTURING THE INVISIBLE

Objects the size of atoms and simple molecules, and even DNA and proteins, are truly invisible because, even with the aid of the highest-magnifying light microscope, our eyes can't see them. Although science does have other powerful ways of finding out what very small things "look" like, nobody really sees details of molecular structures exactly. Thus we have taken liberties in picturing our principal molecular characters in ways that convey clearly what they do.

We depict DNA as a kind of extended Tinker-Toy structure that readily assembles and pulls apart. Proteins — the working molecules of life — are pictured as somewhat human-like little characters. This distinguishes them — things that act — from other molecules, things that are acted upon. We don't imply that proteins are like people in any other ways — except, perhaps, for a certain obsessive tendency to do the same things over and over again. The protein's affable but blank expression should convey the idea.

Sixteen
Things
You
Should
Know
About
Life

Energy Flow Through Life

Why Life Must Come From L

DNA Packaging

Community Energy

Sugar Burning

The Double H

Sugar Making

Enzymes

Energy Molecules

Replication and Repair

Informa Chains

A Four-Letter Alphabet

Chemical Bonds

PATTERNS   ENERGY   INFORMATIO

YOUR ITINERARY — A MAP OF THIS BOOK

The first chapter — "Patterns" — offers a panorama of some of life's key features designed to focus your thinking and whet your appetite. Many of the questions it raises will be answered by the time you've finished the book.

Life sustains itself by converting sunlight into energy. The story of this flow is the subject of Chapter 2, "Energy."

You may find it useful, in starting out, to think of "Information," the subject of Chapter 3, as a primer on the "know-how" for life, written out in the chemical language of DNA, and stashed inside each cell of a living organism. Information, in turn, codes for life's "Machinery," discussed in Chapter 4 — protein molecules that do all life's jobs, including constructing themselves.

Energy, information, and machinery would lead nowhere without some means by which cells can regulate rates of chemical reactions, minimize waste, promote efficiency, and ensure

Cultural Evolution

e Unity of Biology

Ecology Loops

Emergent Patterns

Natural Selection

Pumping Iron

Neural Feedback

Embryo Development

Protein Variety

Chemotaxis

Dividing Contracting Migrating & Dying

Evolving Intelligence

Molecular Signals

Cumulative Changes

Geographic Isolation

Protein Assembly

Circular Information

Folded Chains

Allostery

Cascades

Symbiosis

Sex

Mutations

Gene Switches

Amino Acids

Jumping Genes & Viruses

Origins

ACHINERY FEEDBACK COMMUNITY EVOLUTION

that multiple interlocking processes work harmoniously together to promote the welfare of the whole. This is the role of life's system for coordination and control — which we call "Feedback," the focus of Chapter 5.

All this has to do with what individual cells need to be alive. Chapter 6, "Community," examines principles governing how cells interact with each other in multicellular organisms and, particularly, how a single cell — an egg — becomes a multicellular individual.

Having examined the what and the how of life, we consider the why: why are living things the way they are? As information passes from generation to generation over vast stretches of time, it changes and, inevitably, modifies life's machinery. That machinery, the means by which every organism makes contact with the surrounding world, determines the organism's fate and, consequently, the fate of the information within it. Chapter 7 addresses the theme that knits all of biology together into a comprehensible whole — "Evolution."

Chapter 1

PATTERNS

Sixteen Things You Should Know About Life

To see life as a whole — to observe what all life has in common — requires a shift in the way we normally look at things. We must look beyond the individual insect or tree or flower and seek a more panoramic perspective. We need to think as much about process as we do about structure. From this expanded viewpoint, we can see life in terms of patterns and rules. Using these rules, life builds, organizes, recycles, and re-creates itself.

Here we describe sixteen of life's patterns. Most apply to the smallest organisms and their molecular parts as well as to the most complex of us. We make no claim that our list is definitive. We simply invite the reader to think about life from the standpoint of not just what makes each living thing unique and different, but also what it is that unites us all.

The Sixteen Patterns:

1. *Life Builds from the Bottom Up*
2. *Life Assembles Itself into Chains*
3. *Life Needs an Inside and an Outside*
4. *Life Uses a Few Themes to Generate Many Variations*
5. *Life Organizes with Information*
6. *Life Encourages Variety by Reshuffling Information*
7. *Life Creates with Mistakes*
8. *Life Occurs in Water*
9. *Life Runs on Sugar*
10. *Life Works in Cycles*
11. *Life Recycles Everything It Uses*
12. *Life Maintains Itself by Turnover*
13. *Life Tends to Optimize Rather Than Maximize*
14. *Life Is Opportunistic*
15. *Life Competes Within a Cooperative Framework*
16. *Life Is Interconnected and Interdependent*

1

1. Life Builds from the Bottom Up

The Influence of Small Things

*"Each living creature must be looked at as a microcosm —
a little universe formed of a host of self-propagating organisms,
inconceivably minute and as numerous as the stars in the heaven."*
— *Charles Darwin*

Early debate about evolution centered around the then-horrifying notion that humans and apes had a common ancestor. But Darwin's idea had far more radical implications: Every individual is a colony of smaller individuals (cells), which are in turn made up of smaller nonliving bits. Further, these smaller bits were the first to develop in our evolutionary history. They slowly began to incorporate themselves into cells, which later assembled into multicellular organisms. Our ancestors were microscopic, wriggling, squirming creatures similar to what we now call bacteria, whose own ancestors were bits of self-replicating molecules.

Before a single plant or animal appeared on the planet, bacteria invented all of life's essential chemical systems. They transformed the earth's atmosphere, developed a way to get energy from the sun, devised the first bioelectrical systems, invented sex and locomotion, worked out the genetic machinery, and learned how to merge and organize into new and higher collectives. These are ancestors to be proud of!

Given the complexity of the tasks above, we can see why the first multicellular organisms did not appear until the most recent one-eighth of life's duration on earth. So we exist as "corporate elaborations" — composite communities of cells built out of the accomplishments of our one-celled forebears.

Cooperating Communities of Cells

Small communities of cells — like the taste buds on our tongues — work together as an army of specialists. They create a unique structure, with nerve connections to our brain, that allows us to taste the world around us. (The picture at right represents an enlargement of the human tongue.)

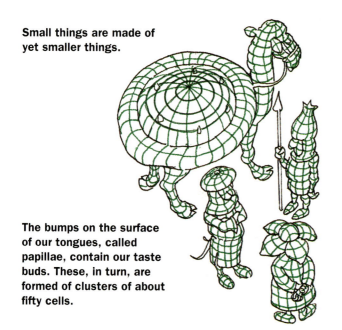

Small things are made of yet smaller things.

The bumps on the surface of our tongues, called papillae, contain our taste buds. These, in turn, are formed of clusters of about fifty cells.

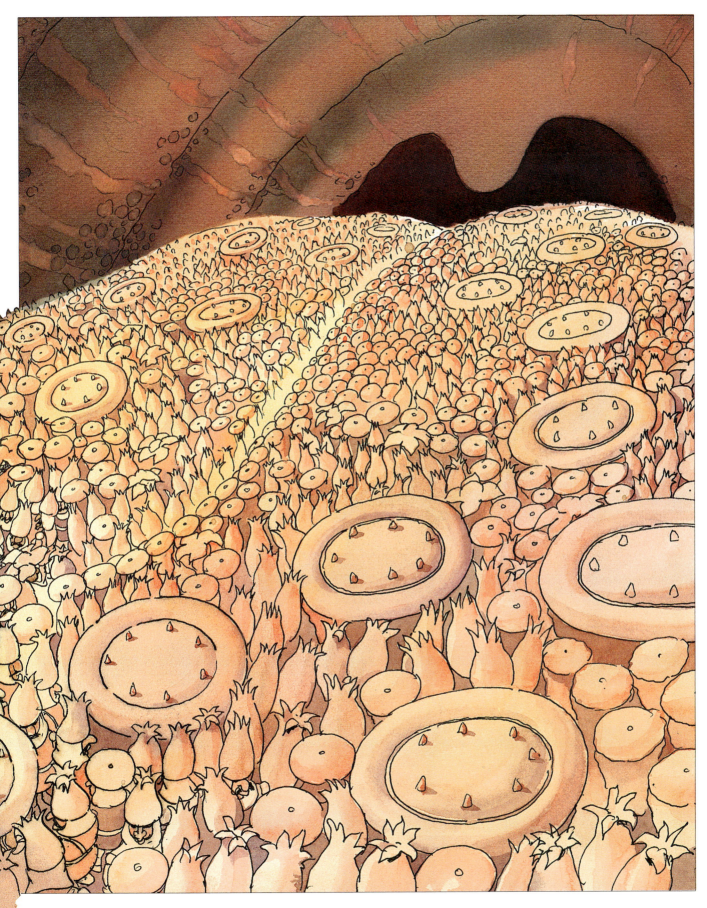

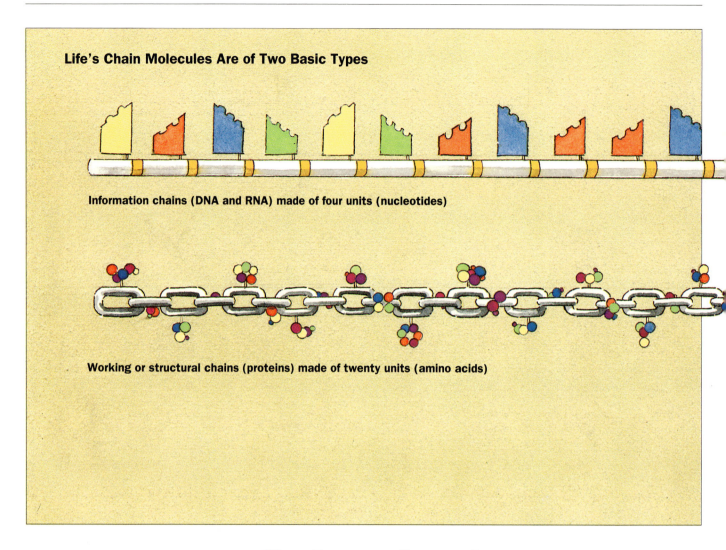

Life's Chain Molecules Are of Two Basic Types

Information chains (DNA and RNA) made of four units (nucleotides)

Working or structural chains (proteins) made of twenty units (amino acids)

Note: The terms DNA, RNA, nucleotides, proteins, and amino acids are explained in Chapters 3 and 4.

When Difference Becomes Information

At the level of its smallest relevant components, life has adopted the chain as its organizing principle. Chains are made of simple units linked together in long, flexible strands. In an ordinary chain, the links are all the same. In contrast, life's chains are made of molecules containing *different* links. In this respect, the links are the alphabet of life. Letters, in appropriate order, form meaningful words, sentences, paragraphs. Similarly, the sequence of individual links in a chain molecule conveys information.

Chain molecules fall into two main classes: information chains, which store and transmit information, and working chains, which carry out the business of living. The two kinds of chains work together in a cooperative loop: Information chains provide the genetic prescription or recipe that is translated into working chains; these in turn make it possible to copy the information chains so they may be passed on to the next generation.

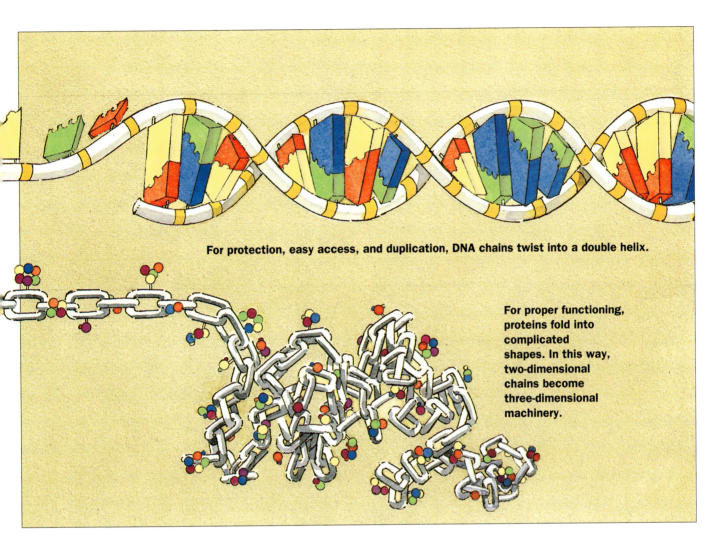

For protection, easy access, and duplication, DNA chains twist into a double helix.

For proper functioning, proteins fold into complicated shapes. In this way, two-dimensional chains become three-dimensional machinery.

A chain of uniform links is simply a chain —

but a chain of different links can carry information:

■● ●■● ●●● ■■● ●■ ●● ● ●●● ●■

Morse code is a chain of two units (dots and dashes),

IOOIIOOOIIIIOIOOOIIOOOIOIIIIOOOIIOOIOOIOI

computer language is also a chain of two units (ones and zeros), and

Now is the winter of our disco

an English sentence is a chain of twenty-six units (letters).

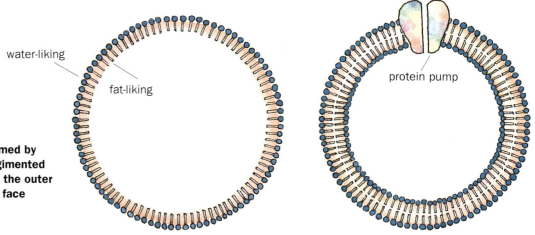

water-liking

fat-liking

protein pump

A Cell Membrane

Cellular membranes are formed by combining two layers of regimented phospholipid molecules. On the outer row, the water-liking heads face outward toward the watery surroundings.

On the inner row, the water-liking heads face toward the inside of the cell. The two rows effectively isolate the inner environment. Protein pumps, like the one shown at top, move molecules in and out.

HEADS OUT — TAILS IN

When danger threatens, musk oxen gather in a circle — heads and horns to the outside, tails to the inside — sheltering their vulnerable calves in the center. This circle of protection illustrates one of life's most fundamental organizing principles — a difference between in and out. Life's chemicals need to be kept close together — concentrated — so that they can meet and react readily. The inner environment needs a saltiness, acidity, temperature, etc., different from the outside. These differences are maintained by some form of protective barrier, e.g., a baby's skin, a clam's shell, or a cell's membrane.

The membranes surrounding each of our cells behave something like the threatened musk oxen. The constituent fat molecules have a water-liking head and a fat-liking tail. Heads face outside toward the watery environment beyond the cell; tails face inward. Since the inside of a cell also has a watery environment, a second row of fat molecules aligns itself tail-to-tail with the outer layer, heads facing inward. With this protective structure creating an inside and an outside, plus several pumps embedded in the membrane to move materials in and waste out, life can do its work.

Larger "Membranes"

Bark safeguards the living part of the trunk (usually the outermost ring) from insects, disease, and harsh weather.

The atmosphere helps regulate the earth's temperature as it protects life from the sun's harmful ultraviolet rays.

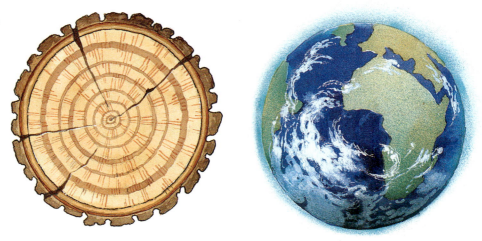

Variations on a Theme

The beetle, with some 300,000 separate species (the world's most numerous order), displays every imaginable color, decorative motif, and proportional distribution of body parts — yet the pattern of relationships that makes the species all beetles is constant.

The Inward Similarity of Outward Diversity

Life hangs on to what works. At the same time, it explores and tinkers. This restless combination leads to a vast array of unique living creatures based on a considerably smaller number of underlying patterns and rules. For example, when cells divide and grow, they do so in a mere handful of ways. New cells can form concentric rings, as they do in tree trunks and animal teeth. They can form spirals, as in snails' shells and rams' horns; radials, as in flowers and starfish; or branches, as in bushes, lungs, and blood vessels. Organisms may display several combinations of these growth patterns, and the scale can vary; but for all life's diversity, few other growth patterns exist.

Life, in striving for the most economical use of space, borrows mathematical rules. For instance, count the branches coming off a stem for a given number of full turns around the stem, and with surprising consistency the numbers of turns and branches relate to each other as in the series 1 1 2 3 5 8 13 21... — the so-called Fibonacci series — in which each successive number is the sum of the two preceding it. Thus, in a pine cone, there are thirteen scales for every eight turns. Similar patterns occur in the spirals of florets in sunflowers and daisies, the sections of the chambered nautilus, even the branchings of the bronchial tubes in our lungs. Such similarities in pattern give us some insight into how simple rules, used in different contexts, can produce great variety. From few notes, nature creates many symphonies.

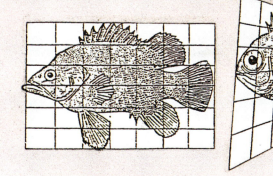

Different Proportions — The Same Pattern

Placing these varied fish species within a "stretchable" grid demonstrates that their differences in shape are a matter of proportion. The fundamental pattern is the same.

MAKING THE PARTS THAT MAKE THE WHOLE

The business of living requires a lot of information. An organism needs to know how to maintain a constant temperature, how to replace worn-out parts, how to defend against invaders, how to get energy out of food, and so on. It has been estimated that the information a human being needs for all of his or her functions would fill up 1,500 encyclopedias. It might be many times greater than that but for a strategy life has developed in storing only a certain kind of information. The nature of this information might be best understood by the following analogy: Suppose you decided to build a complex robot requiring millions of individually hand-crafted working parts. Presumably, this task would require instructions for the making of each part, plus instructions for the overall assembly, as well as operating instructions. But now imagine that you had another option: You could acquire the instructions to make several thousand tiny sub-robots, each of which knew how to fabricate one stage of one of the parts. And by working collectively, these sub-robots could assemble and operate the entire robot. In other words, an extraordinarily complex robot would result from complicated interactions among many sub-robots, each of which performs a relatively simple task.

This is the kind of information that life stores in its DNA — in its genes. Genes contain no information on maintaining temperature, defending against invaders, decorating a home, choosing a mate, etc. They contain only information on how (and when) to make proteins. The rest is up to the proteins.

By itself, a sub-robot could never make a complex working part.

For a team of specialists, however, each completing a single step, the task becomes manageable.

MIXING INSTRUCTIONS

Nature creates new combinations by exchanging information. The earliest life forms, simple bacteria-like organisms, found a way to inject bits of information into each other — a primitive form of sex. Over time, life acquired the ability to exchange ever-larger chunks of information, thus evolving sexual reproduction, which is a more elaborate form of information reshuffling.

Card mixing illustrates an important principle: A very large number of combinations can result from a small number of variables. Shuffling two 52-card decks together produces 2.5×10^{29} 52-card combinations — a staggering number (25 followed by 28 zeros). But now consider what gene mixing can generate. Humans have 50,000 genes, of which roughly 3,000 are non-matched pairs. When we make our sperm or egg, we shuffle our genes together and then cut the deck — divide the genes into two equal pools. Then, on mating, we combine the two halves. The number of possible combinations that any couple can produce is 10^{3000} Compare this with the estimated 10^{80} atoms in the known universe, and you get an idea of the genetic basis of biological individuality.

Mixing Information

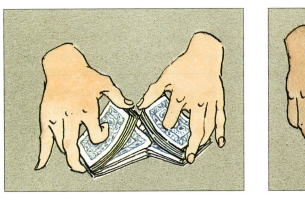

1. Take two decks of cards and shuffle each separately.

 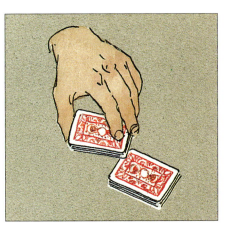

2. Then take half the cards from each deck...

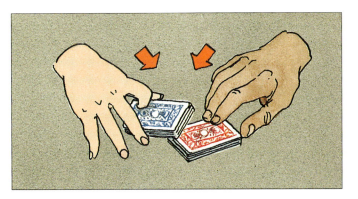

3. ...and shuffle these together.

Size and Surface

Wrinkles and bumps allowed the elephant's ancestors to get bigger. Increasing surface area also allowed organs such as intestines, lungs, and brains to increase their functional size while confined within a limited body space.

A Mistake for One Organism Can Be an Advantage for Another

Albinism, a defect in pigmentation, occasionally shows up in many kinds of plants and animals. Most albinos find themselves at a disadvantage in life, since they don't blend into their surroundings, and albino offspring in many species do not survive infancy. Snowy white polar bears, ptarmigans, arctic foxes, and snowshoe hares, however, owe their camouflaging white coloring (and their very existence) to their albino ancestors.

ACCIDENTS ENSURE NOVELTY

When cells reproduce themselves, they first make a copy of the information they carry in their genes. Usually this copy is exact, so the information is transmitted perfectly to the next generation. But every so often, cellular mechanisms make errors in gene sequences — sometimes by only a tiny bit. Miscopying even a single nucleotide in a gene, like dialing a single wrong digit in a phone number, alters the gene sequence and therefore changes the piece of information being transmitted. The altered information shows up in the offspring, usually as a defect. But every once in a while, it shows up as an improvement — something that makes the offspring better adapted for survival than its parents.

As an example, take the elephant. Scientists speculate that its early ancestors were small and smooth-skinned. Imagine a copying error in the distant past that jumbled the instructions for the elephant's skin cells, making them assemble into wrinkly and bumpy patterns. It happens that wrinkly skin provides more surface area than smooth skin, a fact of geometry that came in handy for the elephant. Large animals generally have a problem with overheating. A wrinkled skin exposes more surface to the air or water and thereby cools the animal more efficiently. Thus wrinkled skin helped make it possible for the elephant to grow larger and to enjoy the advantages that come with increased size.

As you come to appreciate the evolutionary role of copying errors, it is apparent that calling them "mistakes" oversimplifies. We may, in a larger context, view them as nature's way of introducing randomness, an essential feature of all creative processes.

8. Life Occurs in Water

The All-Purpose Molecule

Of all the molecules of life, none is so omnipresent as water. Our cells are 70 percent water. Life began in water. When we arose from the sea to become land dwellers, we brought water along with us, within our cells and bathing them. Most of the essential molecules of life dissolve and transport easily in water.

Water participates in all kinds of chemical reactions. Bounded by water-insoluble membranes, cells owe their shape and rigidity to water. And it provides an inexhaustible supply of the hydrogen ions needed for converting the sun's energy into chemical energy.

What is it about water that makes it so special? The key is its polarity. Composed of a single oxygen atom sharing electrons with two hydrogen atoms — like a head wearing a pair of Mickey Mouse ears — a water molecule looks quite ordinary. While the molecule's overall electric charge is neutral, the oxygen tends to pull negatively charged electrons toward it, leaving the hydrogen "ears" slightly positively charged relative to the more negative oxygen "head." Since most of life's important molecules are also electrically charged, they slip easily through water's charges, becoming soluble and accessible for chemical reactions.

Furthermore, an ear of one water molecule will form weak bonds with the head of another and vice versa, so that water molecules continuously stick and unstick to each other, thus forming dynamic, evanescent lattices. This self-embracing quality of water accounts for its tendency to remain liquid when most other substances with molecules its size are gases.

Luckily for us, water has the unusual property of expanding when it freezes, so that the less dense ice floats. This provides an insulating layer that prevents further freezing of our lakes, rivers, and oceans. If water were like most natural materials, whose solid state is denser than their liquid state, ice would sink, and bodies of water in colder climates would freeze solid, making life untenable.

The most abundant fluid on earth is, happily, the one most suited for encouraging living chemistry.

Water's specialness is due to its molecular structure. The two hydrogens (Mickey's ears) have a positive charge, the oxygen, a negative charge...

...this polarity enables water to form lattices, giving it high viscosity and high surface tension (i.e., its "wetness").

A MOLECULE TO BURN

Sugars are simple, energy-packed chains of three to seven carbon atoms festooned with hydrogens and oxygens. Life's central sugar is the six-carbon glucose. It is the fuel that drives the engine of life and the basic material from which much of life is constructed. Each year, plants, marine algae, and certain kinds of bacteria convert 100 billion tons of atmospheric carbon dioxide (CO_2) and hydrogens extracted from water (H_2O) into sugar — using energy from sunlight in a process called photosynthesis. The waste product of this massive conversion is oxygen.

Plants and algae and bacteria and animals all "burn" sugar. That is, inside their cells they transform the energy in sugar's chemical bonds into an especially potent form of chemical energy — adenosine triphosphate, or ATP. In this living combustion process, called respiration, sugar's carbons and oxygens are discarded as CO_2 and its hydrogens are linked to oxygen and discarded as H_2O. Thus the very substance of life materializes from air and finds its way back to air. The constantly generated ATP powers all life's work, such as moving, breathing, laughing, etc. Sugar also serves as the starting material for the assembly of the simple molecules — amino acids, nucleotides — from which large molecules are assembled.

Several hundred million years ago, enormous quantities of the remains of trees, plants, animals, and bacteria were buried deep in the earth, subjected to intense heat and pressure, and transformed into coal, petroleum, and natural gas. Much of this material was initially chains of sugar molecules — cellulose and other related chain molecules. So sugar reemerges as the basic ingredient of the fuels that drive the engines of civilization.

Plants produce and store sugar for self-consumption. Animals eat the plants or prey on those that do. Bacteria consume the bodies of all. Thus sugar percolates throughout life.

Glucose, life's key sugar molecule, is broken down — metabolized — by living cells, and its parts used to make life's essential molecules.

GLUCOSE

ENERGY

INFORMATION

MATERIALS

Each year terrestrial and marine plants make enough glucose to fill a freight train 30 million miles long.

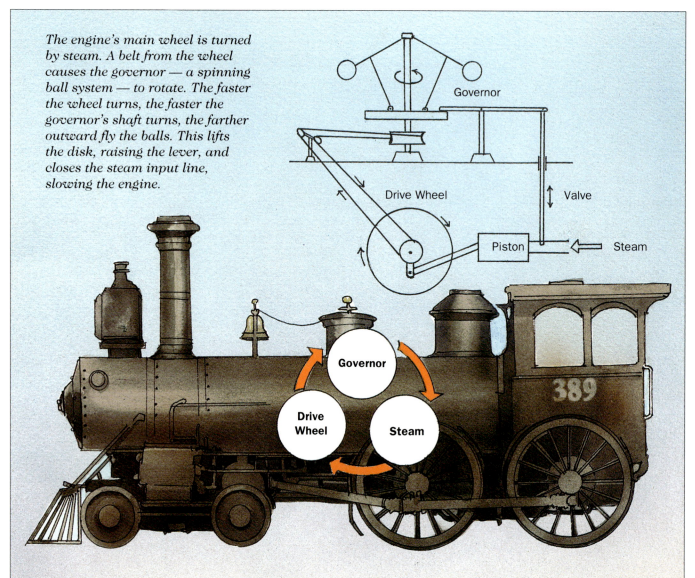

The engine's main wheel is turned by steam. A belt from the wheel causes the governor — a spinning ball system — to rotate. The faster the wheel turns, the faster the governor's shaft turns, the farther outward fly the balls. This lifts the disk, raising the lever, and closes the steam input line, slowing the engine.

Governor

Drive Wheel

Valve

Piston

Steam

Governor

Drive Wheel

Steam

Circular Control

In this simplified steam engine, a fire heats water, making steam, which activates a piston, which turns the engine's drive wheel, which spins the governor, which controls the steam supply. Such a three-component loop passes information from part to part so that the engine is able to self-correct by way of the governor.

A similar self-correcting system comes into play when a protein makes a chemical product. Each protein performs a simple task (e.g., adds a part) in assembly-line fashion. The circular arrangement allows the initial protein to keep track of the overall output. As products either pile up or become scarce, it adjusts the speed of the overall operation. (How it does this is discussed beginning on page 125.)

A Circular Flow of Information

Life loves loops. Most biological processes, even those with very complicated pathways, wind up back where they started. The circulation of blood, the beat of the heart, the nervous system's sensing and responding, menstruation, migration, mating, energy production and consumption, the cycle of birth and death — all have the habit of looping back for a new start.

Loops tame uncontrolled events. One-way processes, given sufficient energy and materials, tend to "run away," to go faster and faster unless they are inhibited or restrained. The steam engine with a governor illustrates the principle: As steam pressure rises, the engine goes faster. The governor, consisting of two rotating arms that lift higher as its shaft spins faster, progressively reduces the steam input; the engine slows; the governor slows; the steam input increases; the engine speeds up. Thus information courses around the circuit to produce action in the opposite direction. The system self-corrects; the parts self-adjust. If such self-generated restraints and inducements occur in small steps, the overall system appears to maintain itself in a steady state.

Every biological circuit, whether a sequence of proteins in the act of consuming a sugar molecule or a complex ecosystem exchanging material and energy, exhibits self-correcting tendencies like those of the steam engine.

Information flows around the circuit and feeds back to the starting point, making necessary adjustments along the way. It's easier to understand how molecular systems assemble into complicated, apparently purposeful organisms when we look at events in terms of multilayered loops of control and creation — and substitute the term "self-correcting" for "purposeful."

Self-Correcting Maneuvers

As an owl tracks a fleeing mouse, she quickly translates the mouse's zigzags into movements of her wings and tail. The owl gets her dinner by maintaining a feedback loop between her eyes, brain, wing and tail muscles and the mouse's movements.

1

2

3

11. LIFE RECYCLES EVERYTHING IT USES

A CIRCULAR FLOW OF MATERIALS

For every molecule that life makes or uses…

…there exists an enzyme somewhere to break it down.

We humans are unique among animals; we leave behind us a trail of accumulating, unusable products. Everywhere else in the living world, intake and output are balanced, and one organism's waste is another's food or building materials. Waste from a cow circulates from bacteria to soil, to earthworms, to grass, and back to the cow. Crabs need calcium, which they normally get from the ocean, to build their shells. Land crabs, lacking an ocean source, extract calcium from their own shells before discarding them during molting. Hermit crabs save energy by moving into shells cast off by other species, trading up when the shell gets too small.

At the molecular level, key atoms pass from molecule to molecule in a succession of small steps. The end product of one process becomes the starting point of another, the whole train of events bending around into a circle. One creature's "exhale" becomes another's "inhale." Oxygen, dumped by plants as a waste product of photosynthesis, becomes an essential key to combustion in animals' respiration. And the carbon dioxide waste that animals excrete is eagerly taken up by plants for sugar-making. From the standpoint of the whole ecosystem, these interchanges occur so smoothly that the distinction between production and consumption, and between waste and nutrient, disappears.

Each generation of living things depends on the chemicals released by the generations that have preceded it.

In a continuous cycle, plants and animals exchange the chemicals necessary for energy and building materials.

CARBON DIOXIDE

OXYGEN

SUGAR

NITROGEN

23

Put It Together — Take It Apart

Consider the following dilemma. To exist, life requires organization. Organization requires energy. Life's complex molecules have lots of energy in the bonds that hold them together, but high-energy bonds don't hold together indefinitely. They tend to fall apart — dissipate. Now, a system that is unstable when it's organized has a problem. How can it avoid inevitable breakdown? Living systems have answered this question with an ingenious strategy. Day in and day out, round the clock, organisms routinely take apart their own perfectly good working molecules and then reassemble them. Each day about 7 percent of your own molecules are "turned over." That means virtually 100 percent have "turned over" in about two weeks. In this way, no molecule lingers in your system long enough to "unintentionally" dissipate.

Turnover also provides flexibility. A change in the environment often calls for a switch in proteins. New proteins can be made from disassembled old ones.

In turnover we can sense life's need for a continuous "flow-through" of energy. A high-information/high-energy state must be dynamically maintained by the ceaseless building and destroying, ordering and disordering, of life's parts.

Keeping a living system in a state of high organization necessitates the continuous building and destroying of its parts.

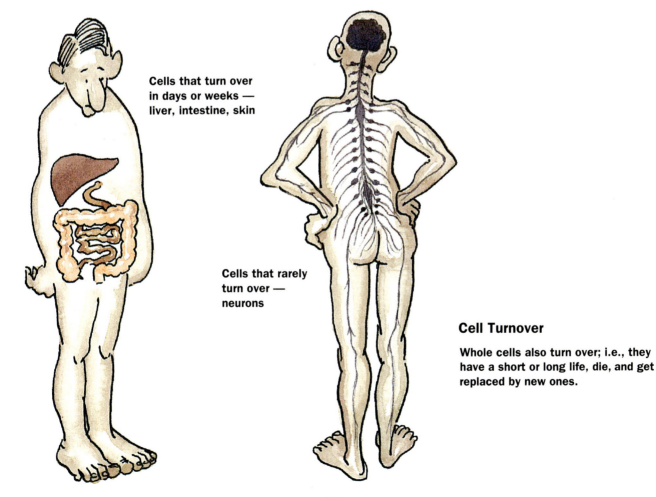

Cells that turn over in days or weeks — liver, intestine, skin

Cells that rarely turn over — neurons

Cell Turnover

Whole cells also turn over; i.e., they have a short or long life, die, and get replaced by new ones.

13. Life Tends to Optimize Rather Than Maximize

When Less Is Better

To optimize means to achieve just the right amount — a value in the middle range between too much and too little. Too much or too little sugar in the blood will kill. Everyone needs calcium and iron, but too much is toxic. The rule of optimization generally holds true for minerals, vitamins, and other nutrients the body requires, as well as for behaviors such as exercise and sleep.

At the molecular level, life operates elaborate signaling and management systems to maintain optimum levels. Certain proteins have the ability to regulate precisely concentrations of essential chemicals, shutting down production when optimum quantities have been reached, starting up again when concentrations fall below critical levels.

At the level of the organism, optimizing is an intricate dance involving many interacting parts and values. Deer antlers require an optimum mix of strength, shock absorption, weight, and growing ability (since they must be regrown every year). A change in any one of these variables might adversely affect the others. Something that might make the antlers stronger, like a higher mineral content, might also make them heavier or unable to grow quickly enough. Thus, maximizing any single value (i.e., pushing it to the extreme) tends to reduce flexibility in the overall system, so that it may not be able to adapt to adverse environmental change.

Maximizing can be seen as a form of addiction. Occasionally organisms may drift from optimizing into maximizing — from adaptation into addiction. Humans exhibit addictive tendencies when trying to maximize such values as wealth, pleasure, security, and power. In restoring optimal balance, we might well take note of nature's dictum: Too much of a good thing is not a good thing.

There is, however, one value that life can be said to maximize. Every organism has as its most elemental goal the transfer of its genetic information to the next generation. In this sense, all optimizing of function aims at this ultimate maximization — the survival of DNA.

Maximizing to Extinction?

The odd positioning (facing forward) and sheer massiveness (up to twelve feet across) of the Irish elk's antlers suggest they were used for display to attract females, rather than for combat. But in the face of major environmental change — such as the heavy growth of forests — "oversized" antlers might well have contributed to the species' disappearance.

14. Life Is Opportunistic

Making the Most of What Is

A rotting tree on the forest floor may look like life at a dead end. In actuality, it marks the beginning of an explosive new stage — more varied and bustling than when the tree was alive. Early on, mosses and lichens establish themselves on the decaying surface. Carpenter ants, beetles, and mites initiate a succession of invasions by channeling through the rotting wood. Fungi, roots, and microbes follow these paths. They in turn become food for grazing insects. Spiders feed on the grazers. Roots of seedling trees and shrubs take hold in the emerging humus as moles and shrews burrow through the soft wood to feed on the newly grown mushrooms and truffles.

The "living dead" tree illustrates not only life's tenacity, but also life's universal tendency to "make do" with whatever is available in its surroundings. Because of this habit, life flourishes even in the world's harshest places. In Africa's Namib Desert, surface temperatures soar to 150°F, and rain may not fall for three or four years at a stretch. Few plants can survive, yet just under the barren sand live a host of insects, spiders, and reptiles — even a mammal or two. The smallest creatures get moisture from wisps of fog and nutrients from tiny bits of plant and animal detritus blowing across the sands. The larger creatures live on the smaller.

In the arctic ice, 100-year-old lichens grow in temperatures of −11°F. Some antarctic fish have a natural antifreeze running through their veins, enabling them to thrive where others would perish. Tubeworms live in darkness 8,000 feet underwater, depending on minerals streaming from hot water vents on the ocean floor. The world's champion adapters, bacteria, can live virtually anywhere — from near-boiling sulfur springs to the acid guts of termites. And so on.

Together, the genetic code and the protein structure of all living things permit a marvelous flexibility. Hence, life forms are opportunists. Opportunists don't wait around for the right conditions. They adapt to what is, and they make use of whatever they find around them.

Self-Burial

To avoid winter's harsh dry winds, the mescaxl cactus withdraws completely into the ground.

Growing Toward Darkness

In order to find a tree to climb, the monstera vine must first grow toward darkness. Once it reaches a trunk, it switches strategies and grows toward light.

Adapted to Fire

Resin in the cone of the lodgepole pine prevents the scales from opening. Fire not only melts the resin and releases the seed, it also leaves a fertile bed of ashes in which the seedling can take root.

An Invitation to Sex

With the right odor, pattern, and degree of hairiness, the bee orchid entices the male bee into an attempt at copulation.

Living Stones

Lithops look like stones, which helps them avoid being eaten by foraging animals.

Hollow Leaves

Moisture condenses on the inside of the pitcher plant's leaves and is then carried directly to the roots, which need to be kept wet because they are exposed to the air.

Like Rotting Meat

With an evil smell, the Rafflesia flower (actually a fungus) encourages pollination by flies.

15. Life Competes Within a Cooperative Framework

Strategies for "Fitting In"

1. Every creature acts in its own interests.

2. The living world works through cooperation.

These two statements may appear contradictory; they are not. Creatures are self-interested but not self-destructive. Selfish behavior, pushed to the extreme, usually has unpleasant costs. A dominant animal engaging in too-frequent combat may sustain injuries. A parasite may kill its host and have nowhere to go. These self-defeating strategies generally get weeded out by evolution, so that in the long run most everyone tends to adopt some form of "getting along."

Up close, the world looks competitive. From a little distance, its cooperative aspects emerge. A million sperm compete in a winner-take-all race to fertilize a single egg. Sperm production is cheap. Nature can afford to make lots of them to ensure that one is successful. We don't weep for the 999,999 losers. They were part of the system to ensure fertilization, and they did their job. Something similar occurs with predator/prey relationships. Usually predators can take only the smallest, weakest, or most unhealthy of their prey species, leaving the fittest members to survive and reproduce. This may be seen as being competitive at the individual level, cooperative at the group level. (Although we don't suggest that creatures generally *think* in terms of the group.)

Noncompetitors

Although these wading birds feed side by side, they might as well be on separate planets. Each eats a different diet with its unique bill. The fact that each species occupies its own special niche may be taken as evidence for nature's desire to "get along."

Plants and animals evolved from predator/prey truces among bacteria. The ancestors of chloroplasts and mitochondria (the sugar-making and sugar-burning components of plant and animal cells, respectively) originally acted as small predators, invading larger bacteria. They exploited but did not destroy their host. Such "restrained predation" is a recurring theme in evolution, and in it we see the beginnings of cooperation. In time, the host developed a tolerance for the invaders, and each began to share the other's metabolized products. Eventually they became full-fledged symbionts — i.e., essential to each other's survival. This progressive cooperation set the stage for all higher life forms. The lesson, as biologist Lewis Thomas has stated, is not "Nice guys finish last," but rather "Nice guys last longer."

From Predation to Cooperation

A parasitic mitochondrium invades a larger bacterium.

Many generations later, invader and host begin to share metabolized products.

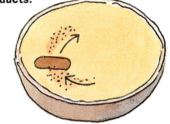

After many more generations, they've come to need each other.

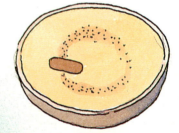

Ritualized Aggression

Animals compete to establish dominance. Such fights rarely result in injury and frequently involve only "displays." This can be seen as cooperative behavior.

Nudibranchs are born defenseless but acquire a protective toxin by eating poisonous anemones and incorporating them into their spines.

Parrot fish deliberately nibble away at the reef while grazing on algae. In the process they excrete calcium as a fine sand. Each fish produces thirty pounds of sand per year, playing an important role in building beaches.

Pink algae use the reef as a secure place to grow. At the same time, they contribute mightily to holding the reef together by secreting a limey "glue."

Crabs encourage sponges to grow on their backs. A good sponge growth discourages octopuses from eating the crab.

Sea squirts carry tiny creatures called nephromyces in their kidney-like organs. Inside the nephromyces live special bacteria. Both the nephromyces and the bacteria appear to be useful in recycling nitrogen for the sea squirt.

A Network of Interactions

The stony coral, a pea-sized animal that resembles a miniature flower, might easily go unnoticed were it not for the tiny limestone cup it secretes for its home site. As the multiplying coral add on their cups, they form vast apartment complexes — the largest life-made structures on earth. Pink algae, taking hold in the crannies, "mortar in" the loose and broken sections with a limey secretion of their own. Turtle grass, sea fans, sponges, and mollusks attach themselves to the reef surface. Moray eels take up residence in the dark crevices. Starfish arrive to feed on the coral, and triton conches feed on the starfish. Hundreds of species of fish — some grazers, some predators — move in, along with crabs, octopuses, shrimp, and sea urchins. Competitive and cooperative relationships emerge.

Damselfish flit with complete immunity among the poisonous tentacles of the large sea anemones. Crabs encourage sponges to grow on their backs as a protection from octopuses. Cleaner fish and shrimp remove parasites from predator fish, even entering their gills and mouths with complete safety. Algae live comfortably inside the coral's cells, and large sponges offer housing to thousands of minute creatures.

Look at the coral reef as a multilevel, integrated system. Ultimately, everything in the reef connects with everything else. The survival of the reef shark is closely tied to the survival of the coral polyp, even though the two may have no contact and certainly no awareness of each other. What survives and evolves are patterns of organization — the organism plus its strategies for making a living and for fitting in. Any successful change of strategy by one organism will create a ripple of adjustments in the reef community. Called coevolution, this is the kind of creative force at work everywhere life has taken hold.

Reef-building coral polyps harbor tiny algae within their cells. The algae promote the coral's growth and receive carbon dioxide and nutrients in exchange.

Cleaner fish live safely in the mouths and gills of larger fish, removing parasites.

Chapter 2

ENERGY

Light to Life

Every day, countless tiny packets of light called photons radiate from the sun, travel 93 million miles across space, and strike the earth. There, the energy of that light is turned into the energy of heat, which stirs the molecules in air, water, sand, and stone. Life thrived in this sunlight-to-heat energy stream, diverting part of its flow into structures that can move, grow, and duplicate themselves. Life accomplished this by finding a way to use the energy of sunlight to make energy-rich molecules, which, in turn, could be used to bond together simple molecules into more complex, long-chain molecules. Think of plant and animal life on earth, then, as an ordered collection of molecules joined by bonds made of captured energy.

This chapter is divided into two equal halves. In the first half, we review some basic chemistry, introduce the key molecular players, and present an overview of life's energy flow. This covers everything you'll need to know to understand any reference to energy throughout the book. If you're interested in more detail, the second half of the chapter (beginning on page 54) shows, in step-by-step sequences, how life captures, stores, and consumes energy. We caution you: this material defies our efforts to simplify — life is, alas, complex.

Plants, animals, and microbes make up a vast cellular carpet spread over the globe. This carpet requires a constant supply of energy from sunlight to maintain itself; it ultimately releases most of that energy as heat.

Sometimes forceful collisions can bond atoms together into molecules...

...theoretically, successive collisions could form a chain of molecules.

A CHAOS OF COLLISIONS

In New York City's Grand Central Station, busy travelers dash about in seeming random fashion on their various missions. Collisions inevitably happen. Imagine that some of these commuters collide so forcefully that they stick together permanently! Now imagine these commuters as atoms, which also bump into one another constantly. When they meet with the right fit and sufficient force, they form a chemical bond — and a molecule is born. Such chemical reactions underlie everything that's happening around us and inside us.

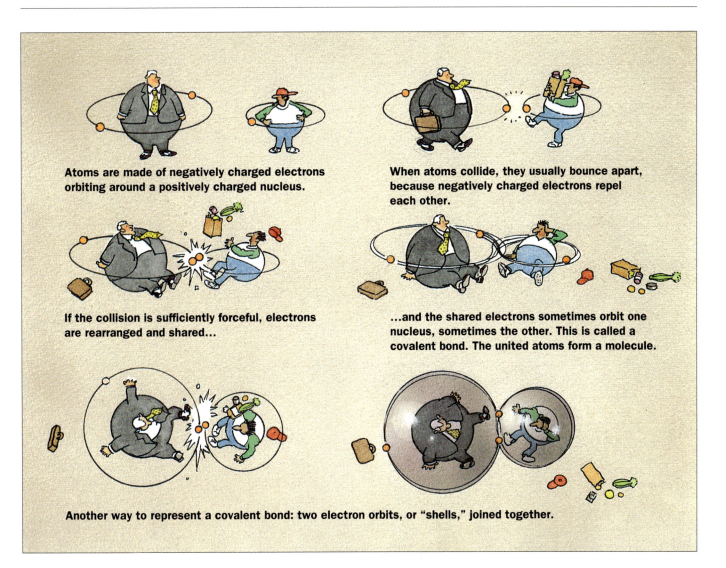

Atoms are made of negatively charged electrons orbiting around a positively charged nucleus.

When atoms collide, they usually bounce apart, because negatively charged electrons repel each other.

If the collision is sufficiently forceful, electrons are rearranged and shared...

...and the shared electrons sometimes orbit one nucleus, sometimes the other. This is called a covalent bond. The united atoms form a molecule.

Another way to represent a covalent bond: two electron orbits, or "shells," joined together.

How Atoms Stick Together

For every atom, the number of positively charged protons in the nucleus equals the number of negatively charged electrons orbiting the nucleus; therefore, overall the atom is neutral. Each kind of atom has a different number of protons in its nucleus — and, consequently, a matching number of electrons in orbits — which accounts for atoms' different sizes and weights. Oxygen, for example, contains eight protons; carbon, six; hydrogen, one. There are over 100 known kinds of atoms in the universe. Only about 20 are important in life.

Let's back up a step, and take a closer look at the atom. It consists of a positively charged nucleus — containing positively charged protons and uncharged neutrons — which is orbited by energetic, fast-moving, negatively charged electrons. When atoms collide, like the Grand Central commuters, their orbiting electrons push them apart, because like charges repel. However, as atoms careen through space, they possess what is called kinetic energy — the energy of motion. If the kinetic energy of two colliding atoms is great enough, it overcomes the repulsion of their electrons and a chemical reaction occurs, causing a rearrangement of electrons and uniting the atoms. Some of the atoms' electrons become *shared* by the two of them, producing what is called a covalent bond. These are strong bonds. They hold life's key atoms — carbon, hydrogen, oxygen, nitrogen, phosphorus, etc. — together in simple molecules, and they join those simple molecules together in chains.

Bonds are also a reservoir of the energy that went into making them. That energy, like fuel, can be put to work in cells to accomplish life's feats of moving, growing, and reproducing.

Sometimes forceful collisions can break bonds apart.

BREAKING BONDS

Back at Grand Central Station, the concourse is now full of "molecules" — human "atoms" stuck together as a result of earlier collisions. In their ongoing mad dash for their trains, these molecules will frequently bump into each other without effect. But, now and then, a collision with more than the usual force and at just the right angle will cause the bonds between molecules to break. When a bond breaks, the shared electrons fall back into the original orbits around the separated atoms, releasing the energy in the bond as heat.

Cells need to be able to break bonds to rearrange molecules in all sorts of ways and to dispose of molecules no longer needed.

The energy is released as heat.

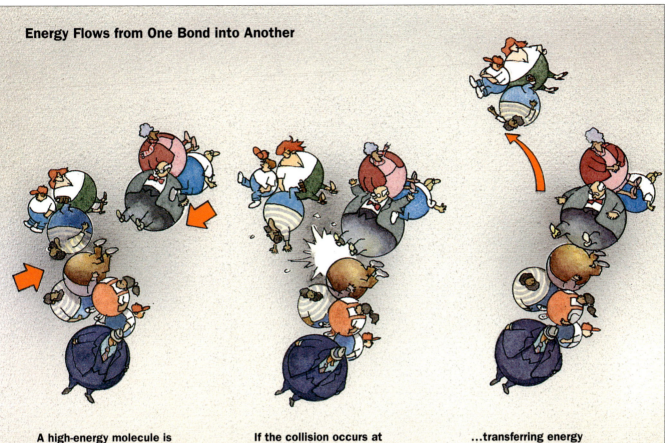

Energy Flows from One Bond into Another

A high-energy molecule is about to collide with the molecule approaching from the right.

If the collision occurs at the right place and at just the right angle, the key bond in the high-energy molecule will break...

...transferring energy to the new molecule and discarding the displaced portion of the high-energy molecule.

Life is possible because of the great variety of molecular combinations. Using mostly carbon, hydrogen, oxygen, nitrogen, phosphorus and sulfur, life fashions all its simple molecules and a near-infinite variety of large chain molecules.

Transferring Energy

Key bonds in certain kinds of molecules can produce an unusual amount of energy. When these high-energy bonds are broken, the energy in them can be *transferred* to other molecules, instead of being lost as heat. That energy is captured and preserved in a new bond between part of the high-energy molecule and the new molecule to which the energy has been transferred. All the important activities of cells, such as constructing and moving, are carried out by large molecules of protein, the worker molecules of life, which manage energy through this kind of transfer. Every time a bird flaps a wing, a maple tree sprouts a branch, or a clam opens its shell, bond energy is being transferred. Indeed, everything that happens in living cells is the result of various combinations of bond-breaking, bond-making, and bond-transfer.

LIFE AND THE LAWS OF ENERGY

— This is order

This is disorder —

It seems strange that in a universe where matter and energy disperse — i.e., run downhill — life congregates and organizes — i.e., runs uphill. This contradiction, symbolized at the right, is more apparent than real, as we explain on the following page.

and things tend to go from order to disorder.

I gotta get organized!

RUNNING UPHILL IN A DOWNHILL UNIVERSE

Incredibly, all the chemical processes of life and, indeed, all the energy and matter in the universe obey two simple laws: the *laws of thermodynamics*. The first law says that energy can be gained or lost in chemical processes — shifted from one form to another — but it can't be created or destroyed. Income and expenditure of energy have to balance. The second law says that energy inevitably disperses, dissipates, scatters — that is, it is transformed from more usable forms such as photons and bonds to a less usable form, namely heat. The tendency of energy to disperse, and of ordered structures to become disordered, is called entropy, and physicists say that the entropy of the universe is increasing.

And this brings up a puzzle. If the universe is dispersing its energy, if things are generally running down, how is it that life seems to be going the other way? Paradoxically, while energy has been spreading out, life appears to have gotten increasingly more ordered and complex over time. How can life build uphill with energy that runs only downhill?

In considering this question, we begin with the basic truth that life never contravenes, outwits, or otherwise gets around the fundamental laws of nature. It simply finds ways of using those laws to its own advantage.

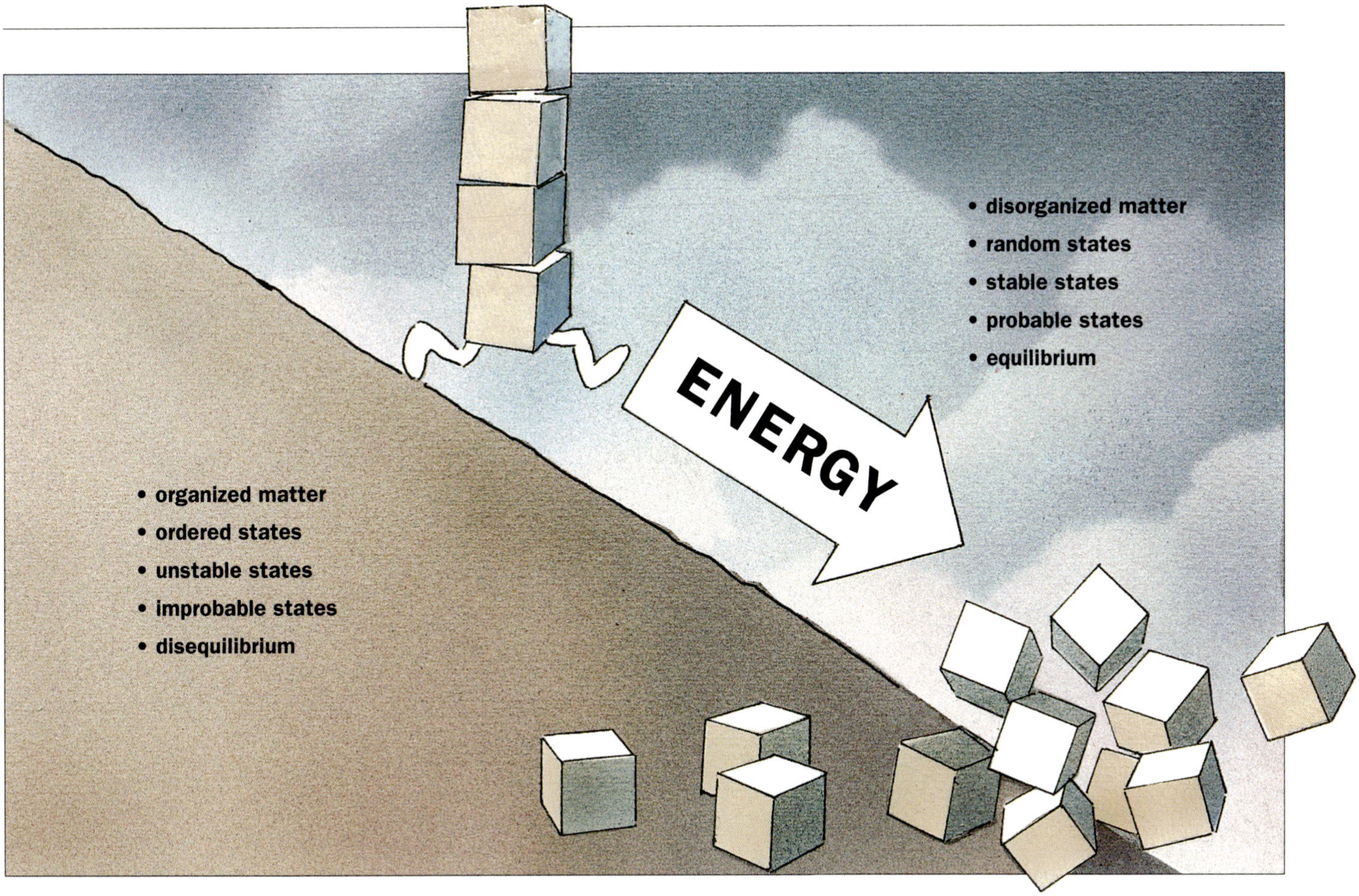

- disorganized matter
- random states
- stable states
- probable states
- equilibrium

ENERGY

- organized matter
- ordered states
- unstable states
- improbable states
- disequilibrium

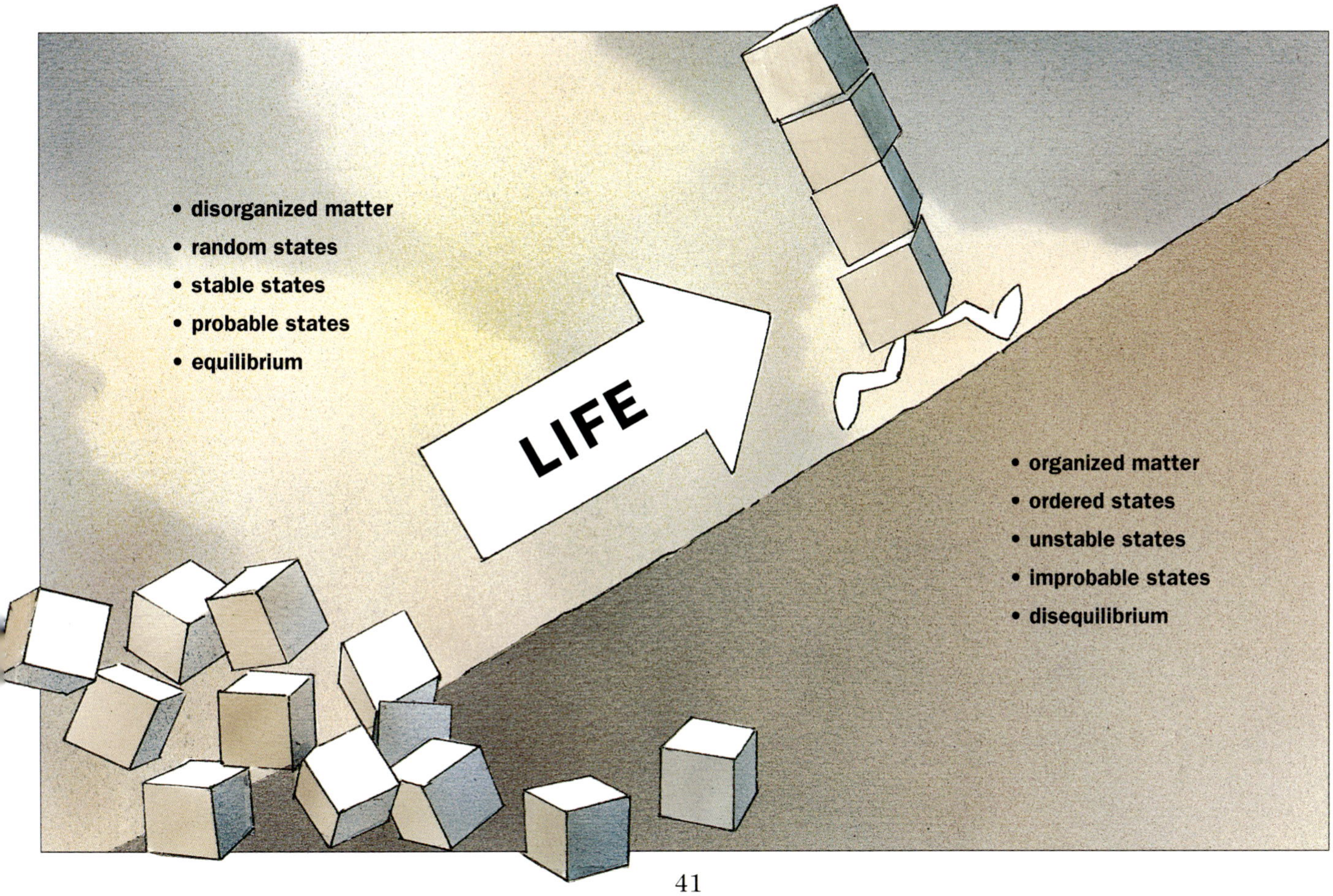

- disorganized matter
- random states
- stable states
- probable states
- equilibrium

LIFE

- organized matter
- ordered states
- unstable states
- improbable states
- disequilibrium

41

THE GOOD NEWS ABOUT THE SECOND LAW OF THERMODYNAMICS

Consider earth's lucky circumstances. Our planet orbits just near enough to the sun to take advantage of its unlimited, steady output of energy without burning up. The sun's light and heat flow over earth and then on into the quiet and cold of outer space, where the temperature is near –273°C — what scientists call absolute zero. This constant flow, described by the second law of thermodynamics, keeps our planet in a comfortable yet energized state in which bond-making, bond-breaking, and energy transfers occur readily. Action and change occur because energy moves toward a more scattered state; once it fully spreads out, nothing can happen: nothing moves, nothing has direction, time itself stops.

Let's take a close-up look at bond-making to see how this dynamic state of affairs works. Each time a bond connecting the simple molecules of life is created, some of the energy put into the bond is used to make it and some is dispersed into the surroundings as heat. In other words, more energy goes into making a bond than actually ends up *in* the bond; the excess is spread out into the surroundings. This seemingly wasteful dispersal of energy as heat, which is demanded by the second law of thermodynamics, has a beneficial effect. Think of it this way: If some of the bond-making energy didn't disperse but stayed nearby, it could readily flow right back and *unmake* the bond. The heat dispersal is necessary to ensure that what gets put together stays together — that the building process is one-way.

Thus, the second law of thermodynamics does not threaten life, but instead guarantees: (1) a steady stream of usable energy dispersed by the sun, and (2) stable molecules with which to build. Running uphill requires no special tricks, but rather a dogged and persistent repairing and rebuilding at the molecular level (like the castle-building crabs at the right).

Some of the energy that goes into the making of the bond is scattered as heat.

This ensures that the bond will be stable enough for the construction work of life. Breaking the bond would require at least as much energy as it took to make it.

A Sand Castle Analogy

A sand castle is a vivid analogy for the effects of entropy. Inevitably, powerful natural forces — waves — will reduce the castle to the random disorder of the sand grains from which it arose.

In the inanimate world, what gets dispersed stays dispersed.

Life can neither circumvent nor otherwise escape the second law of thermodynamics, but it can, for a time, resist the tendency to disperse. Suppose, as a fanciful example, that after each wave, a colony of crabs rushes in and make repairs so feverishly that the castle is completely restored before the next wave.

Of course, crabs don't actually behave this way, but in living systems, proteins perform the job of rebuilding. Their activities require a steady input of energy supplied by the sun and then converted to high-energy bonds. In the animate world, what gets dispersed generally gets rebuilt.

Initially, cream molecules and coffee molecules are separate (as shown in the cutaway section).

Random movement and collisions begin to disperse the cream.

In time, the cream molecules will disperse throughout the coffee.

ENERGY FLOW AND EQUILIBRIUM

Life is a big bag of chemical reactions.

Imagine that you've shrunk to the size of a cell and can watch a chemical reaction take place. A cell is about to put together a bigger molecule out of some smaller molecules. We call the molecules (or atoms) present at the beginning of a chemical reaction *reactants* and the resulting molecules *products*. Remember, when we talk about chemical reactions, we're usually talking about millions of molecules in a confined space constantly rushing around and colliding with each other. The more molecules there are — the more people crowding Grand Central Station — the more collisions there'll be and, therefore, the more likely that chemical rearrangements will happen.

A chemical reaction starts with lots of reactants and no products. Within seconds, millions of reactants get converted into products. As the products pile up, things begin to slow down. Finally, when the energy stored in reactants and products is equalized, no further products accumulate. The molecules have not stopped reacting with each other, however. Collisions continue to convert reactants to products, but now an equal number of collisions convert products back into reactants. When the energy flows as readily backward as forward, no further overall change takes place. This state of affairs is called equilibrium. (The flea-bitten dogs at the right illustrate the principle.) Life generally abhors equilibrium, because that's when cells become inactive and die. By ceaselessly adding reactants and removing products, living cells maintain themselves in far-from-equilibrium conditions.

No Need to Stir in the Cream

The second law of thermodynamics is illustrated by the tendency of cream to disperse in coffee. Once the cream molecules thoroughly disperse, they stay that way. The chance that they'll all float back to the surface is virtually zero. Even though they continue to move and bump into other molecules, they remain more or less evenly dispersed.

How a Dog Shares Its Fleas

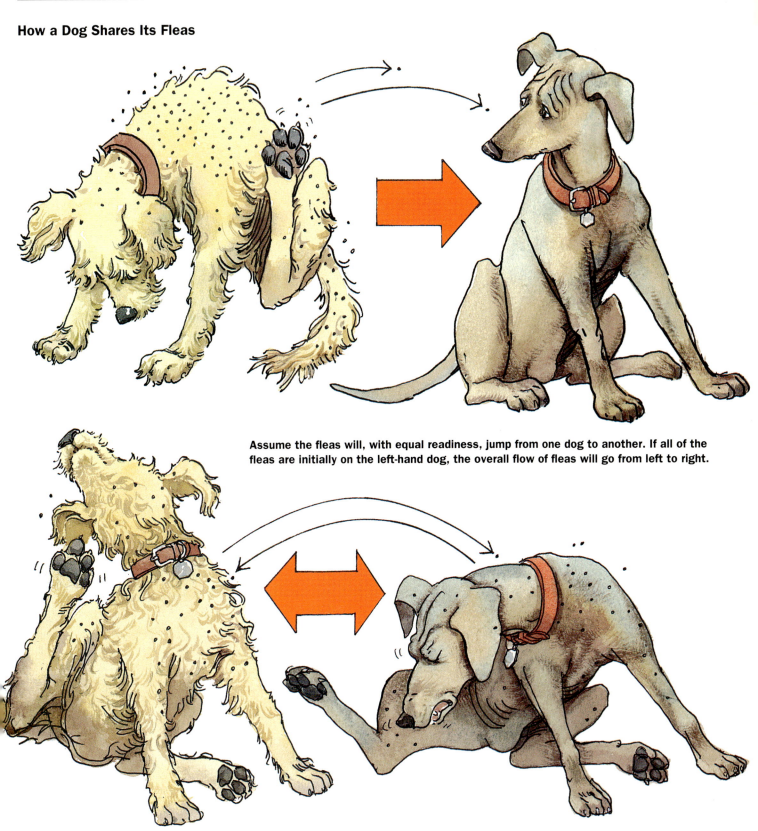

Assume the fleas will, with equal readiness, jump from one dog to another. If all of the fleas are initially on the left-hand dog, the overall flow of fleas will go from left to right.

In time, the fleas will divide themselves equally between the two dogs and *remain equally divided* even though individual fleas will continue to jump back and forth at the same rate as before. This is equilibrium. To keep the fleas *flowing* from left to right, we would have to put more fleas on the left dog or take fleas off the right dog.

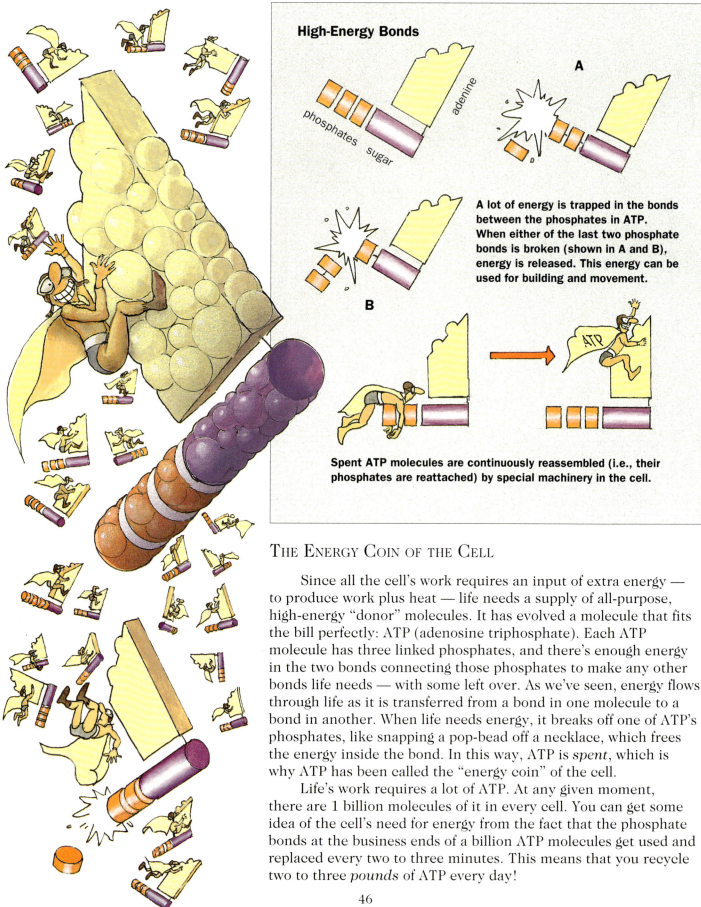

High-Energy Bonds

phosphates sugar adenine

A

A lot of energy is trapped in the bonds between the phosphates in ATP. When either of the last two phosphate bonds is broken (shown in A and B), energy is released. This energy can be used for building and movement.

B

ATP

Spent ATP molecules are continuously reassembled (i.e., their phosphates are reattached) by special machinery in the cell.

THE ENERGY COIN OF THE CELL

Since all the cell's work requires an input of extra energy — to produce work plus heat — life needs a supply of all-purpose, high-energy "donor" molecules. It has evolved a molecule that fits the bill perfectly: ATP (adenosine triphosphate). Each ATP molecule has three linked phosphates, and there's enough energy in the two bonds connecting those phosphates to make any other bonds life needs — with some left over. As we've seen, energy flows through life as it is transferred from a bond in one molecule to a bond in another. When life needs energy, it breaks off one of ATP's phosphates, like snapping a pop-bead off a necklace, which frees the energy inside the bond. In this way, ATP is *spent*, which is why ATP has been called the "energy coin" of the cell.

Life's work requires a lot of ATP. At any given moment, there are 1 billion molecules of it in every cell. You can get some idea of the cell's need for energy from the fact that the phosphate bonds at the business ends of a billion ATP molecules get used and replaced every two to three minutes. This means that you recycle two to three *pounds* of ATP every day!

A Versatile Player: Some of ATP's Jobs

**1. Making information chains
(see page 90)**

**2. Making proteins contract —
as in muscular movement
(see page 106)**

3. Transporting small molecules

cell membrane

protein

protein

4. And also:

Helping to make sugar in
photosynthesis (see page 58)
and bonding molecules together
(see next page).

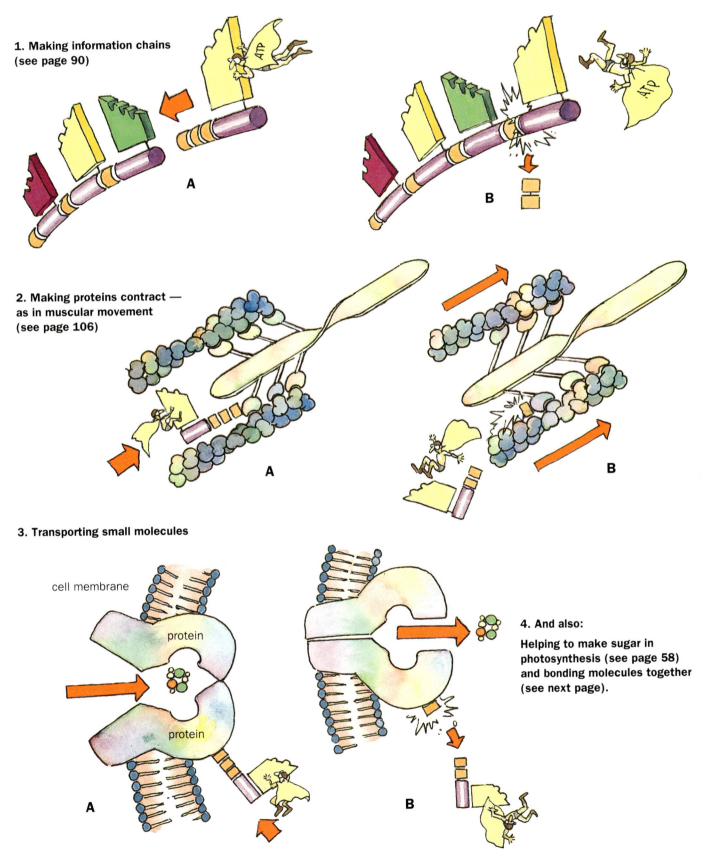

Each Enzyme Has a Specialized Function

Some break molecules apart, some help bond molecules together, some rearrange molecules.

Enzymes have special docking sites for encouraging reactions among small molecules.

ORCHESTRATORS OF CHEMICAL REACTIONS

Life can't get by on energy alone. The simple chemistry of random motion and collision we've seen so far could not begin to maintain life in all its complexity. Things can't be left to chance; life needs a way of making chemical events happen more surely and rapidly. Getting molecules into perfect positions and then pushing them to react is the job of enzymes. Enzymes are catalysts — speeder-uppers and facilitators of chemical reactions. Each enzyme has docking sites on its surface into which simple molecules fit precisely. Once it has a grip on the molecules, the enzyme chemically interacts with them, forcing them to react — in what we might call an aided collision.

We have thousands of different kinds of enzymes in our cells. They are big molecules — hundreds to thousands of times bigger than the simple molecules they work on. They are almost always protein — long chains of simpler molecules (amino acids, see page 4) that twist, bend, and fold themselves into many different shapes, most often resembling gnarled, lumpy potatoes. Their variety and versatility are awesome. They manipulate other molecules (i.e., act as catalysts), regulate production lines, "read" DNA's instructions, receive and react to chemical signals, and more.

How Life Joins "Reluctant" Molecules

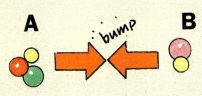

An enzyme and an ATP molecule form a dynamic duo, a sort of Batman and Robin of the cell, working together to accomplish life's tasks of moving and building. Here we see how they solve the problem of joining two reluctant molecules together.

1. Molecules A and B can bump into each other indefinitely but will rarely form a bond. They are "reluctant" partners.

2. An enzyme takes molecule A into its docking site along with an ATP molecule.

3. With careful positioning, the enzyme transfers one of ATP's phosphates to the reluctant molecule.

4. The enzyme then discards the rest of the ATP, which is later rebuilt so that it can provide energy for other reactions.

5. Next, the enzyme takes molecule B into a nearby docking site.

6. Again, with careful positioning, the enzyme breaks the phosphate off molecule A, simultaneously transferring the energy to a bond between A and B.

7. Now the two reluctant molecules are bonded together, and the spent phosphate is discarded.

ENERGY FLOW THROUGH LIFE — A MACRO VIEW

Primary Producers

By far the largest portion of biomass on earth belongs to the photosynthetic organisms: green plants, algae, plankton, and photosynthetic bacteria. These are the sugar-makers.

Herbivores

The largest group of animals are those that directly consume plant sugar. This group includes all of the browsers and grazers, the seed- and fruit-eating birds, most insects, and the ocean's plankton eaters.

Carnivores

This diverse group of predators and scavengers eats the animals that eat the plants. This group includes all members of the cat and dog families, most aquatic mammals, most reptiles, spiders, starfish, even a few plants such as the Venus flytrap.

Decomposers

This group extracts the last bit of energy from the biomass by breaking down the excretions and the dead bodies of the other three levels. Consisting mostly of bacteria and fungi, this group also includes maggots, dung beetles, and earthworms.

FROM PLANTS TO HERBIVORES TO CARNIVORES

Energy flows through each individual and indeed, through the whole carpet of life, one way and downhill. All the energy that enters earth's biosphere eventually goes out again, dispersed into outer space as heat. Along the way, however, energy percolates through several "consumer" levels (symbolized as segregated compartments at the left).

At the first level, green plants (and photosynthetic bacteria) capture the energy from sunlight and put it into the chemical bonds of sugar — life's universal food — in the process called photosynthesis. Plants make sugar for their own use, but they make enough of it to support all other life as well. Herbivores (plant eaters) get their sugar directly from plants, and carnivores (meat eaters) get theirs from the flesh of herbivores. A fourth group, the "decomposers" (i.e., mostly bacteria and fungi) get sugar by breaking down the waste products and dead bodies of the other three groups.

Most of any living creature's energy gets consumed in metabolism — the internal process of building up and breaking down materials. The amount left in its body, as chemical bonds, represents only a small part of the energy that has passed through it. Therefore, the food "chain" in fact looks more like an inverted pyramid, because each successive level extracts only a portion of the energy flowing through it. A square mile of grassland feeds perhaps 100 gazelles, which is about the population needed to sustain a single lion. So each gazelle's take is only about 1 percent of all the grass that is consumed, and the lion gets only about 1 percent of the gazelles (hardly a lion's share!). In a sustainable system, there can never be more lions than gazelles or more gazelles than grass.

Plants, then (and incidentally, bacteria long before plants existed), are the vanguard of life. We animals could only evolve after plant-made sugar (and oxygen, which is a waste product of sugar-making) became plentiful on earth. In fact, we are triply indebted to plants: (1) for our fuel, (2) for the oxygen to burn it, and (3) for saving us from being cooked by the effects of the carbon dioxide produced when we burn that fuel. Carbon dioxide, accumulating in the atmosphere from life and from the industrial processes we have invented, prevents heat from escaping the earth. Plants consume that carbon dioxide in enormous quantities, thereby protecting us from overheating.

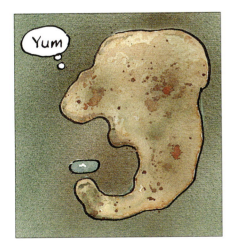

At every level, organisms that cannot convert the sun's energy are dependent on those that can. Here an amoeba eats a photosynthetic bacterium.

From Sugar-Making to Sugar-Burning

Energy flows through the biomass of plants, animals, and microorganisms in complex food chains. At the level where energy is actually captured, transferred, and put to work, a simpler pattern prevails. Incredibly, the entire plant and animal world runs on the work of just two kinds of bacteria-sized organelles within cells. Chloroplasts in plant cells make sugar using the energy of sunlight, in the process called photosynthesis. Mitochondria, in both plant and animal cells, break down (burn) the sugar and make ATP, in the process called *respiration*. So the following general sequence represents the real flow of energy through life: Sunlight —> sugar —> ATP —> heat (released when ATP is used).

The system is actually one step more complicated: In order to make sugar, a chloroplast must first make its own ATP. Its sugar-building enzymes need energy from ATP to work. You may wonder, then, if chloroplasts can make ATP, why do they bother to make sugar? Because sugar provides not only energy, but *building material* as well. As we mentioned in Chapter 1, cells can convert sugar (glucose) into an array of molecules with which to build, notably amino acids to make proteins and nucleotides to make RNA and DNA.

So, if we follow building material through life, rather than energy, we see it flowing in an ever-renewing cycle: Chloroplasts take in the simplest of molecules — carbon dioxide and water — make sugar, and discard oxygen. Mitochondria do the opposite: They take in oxygen and sugar, merge them in burning, and discard water and carbon dioxide. Together, these cellular operations yield a simple and beautiful circularity: water and carbon dioxide in, water and carbon dioxide out, and an almost unimaginable complexity between.

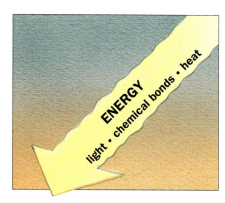

Energy *flows through* life in a one-way stream.

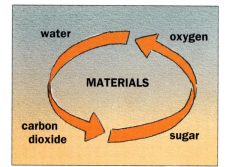

Life's molecules are recycled in a *continuous loop.*

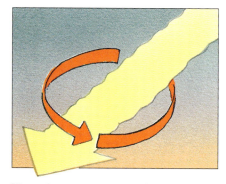

These two processes — *flow-through of energy* and *recycling of materials* — are superimposed in living systems.

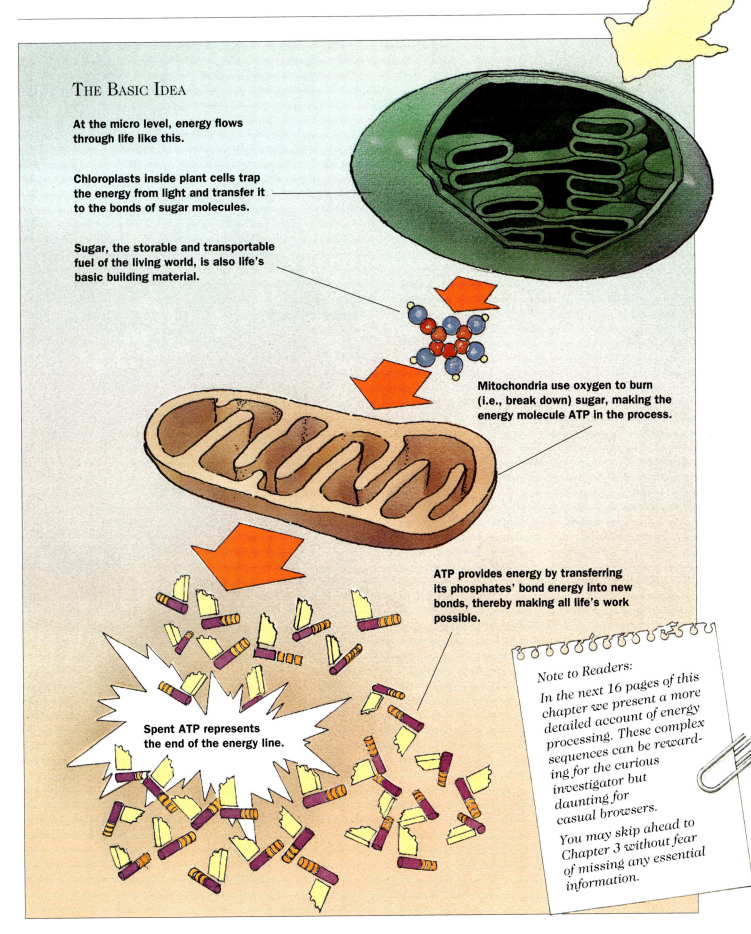

THE BASIC IDEA

At the micro level, energy flows through life like this.

Chloroplasts inside plant cells trap the energy from light and transfer it to the bonds of sugar molecules.

Sugar, the storable and transportable fuel of the living world, is also life's basic building material.

Mitochondria use oxygen to burn (i.e., break down) sugar, making the energy molecule ATP in the process.

ATP provides energy by transferring its phosphates' bond energy into new bonds, thereby making all life's work possible.

Spent ATP represents the end of the energy line.

Note to Readers:

In the next 16 pages of this chapter we present a more detailed account of energy processing. These complex sequences can be rewarding for the curious investigator but daunting for casual browsers.

You may skip ahead to Chapter 3 without fear of missing any essential information.

THE CHLOROPLAST BALLROOM

An overexcited electron flies off one of the dancers...

...and onto an onlooker...

...who, in turn, is energized.

In this way, the energy jumps from dancer to dancer.

THE ELECTRON BOUNCE

Glittering strobe lights animate The Chloroplast Ballroom. As the lights spin and the band breaks into "Sugar Jump," the dancers go wild. Suddenly a bystander inspired by a couple on the floor gets dancing feet. This in turn excites a second bystander to dance and before long a chain reaction takes place, each new dancer energizing the next bystander.

These jitterbugging molecules are demonstrating the initial steps in converting light energy into chemical energy — that is, ATP. Energy waves from light can excite certain electrons in molecules, boosting them into higher-energy orbits. Such energized electrons will actually jump from one molecule to another to another, setting up an electron flow. This leads to the next step — the "Ion Shuffle."

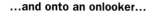

54

HYDROGEN IONS

Hydrogen (symbolized as H), the smallest of all atoms, can readily give up its single electron, leaving behind its positively charged nucleus — a hydrogen ion (symbolized as H+).

THE ION SHUFFLE

Energized female dancers (negatively charged because they have picked up flowing electrons) whirl by males (positively charged hydrogen ions that have lost their electrons) and dance them by a burly bouncer — who grabs each man and throws him into a nearby lounge room. The more the men crowd together, the more desperately they want out. They can only exit through a revolving door, cranking up a machine that assembles ATP.

1. The female dancers team up with male partners — because opposites naturally attract.

3. The more men in the lounge, the more they want out of the crowded room. (Remember the second law of thermodynamics.)

2. A bouncer grabs each man as he dances by and throws him into a lounge. The women leave, worn out from dancing.

4. The only way out is through a revolving door, which spins as they escape...

5. ...operating a machine that reattaches the end phosphate to spent ATP molecules.

A Cross-Section of a Leaf

Shows the chlorophyll-containing cells sandwiched between protective layers of surface cells.

A Single Leaf Cell

Has about 50 chloroplasts — the factories that do the work of producing sugar.

It is estimated that a full-grown, healthy maple tree has about 500 square feet of leaves weighing about 500 pounds. This represents a total chloroplast surface area of about 140 square miles. A single maple can make two tons of sugar on one good sunny day!

IT'S NOT EASY TO MAKE SUGAR

Here's a short-hand version of how green plants make sugar. Energy-filled packets of sunlight (photons) hit chlorophyll molecules in leaves, kicking electrons in those molecules into higher-energy orbits (1). These energetic electrons bounce along a series of chlorophyll molecules (our female jitterbuggers) and onto small carrier molecules (other female dancers) (2). The electrons lost from chlorophyll are replaced by electrons from water, readying the chlorophyll for more action (3). The carriers pick up hydrogen ions (our male partners) and escort them (4) to a protein (our bouncer), which ejects them into a thylakoid sac (the lounge) within the chloroplast (5). The ions, crammed together, force their way out of the sac through a channel in an enzyme (our lounge's revolving door), a process that empowers the enzyme to make ATP (6). The electrons, after getting another energy boost from light (7), finally unite with more hydrogen atoms on a special molecule, NADP, forming highly reactive "hot" hydrogens (8). Finally, a team of enzymes, using the ATP for energy, grabs carbon dioxide molecules from the surrounding air, combines them with the "hot" hydrogens, and links them together to produce sugar (9).

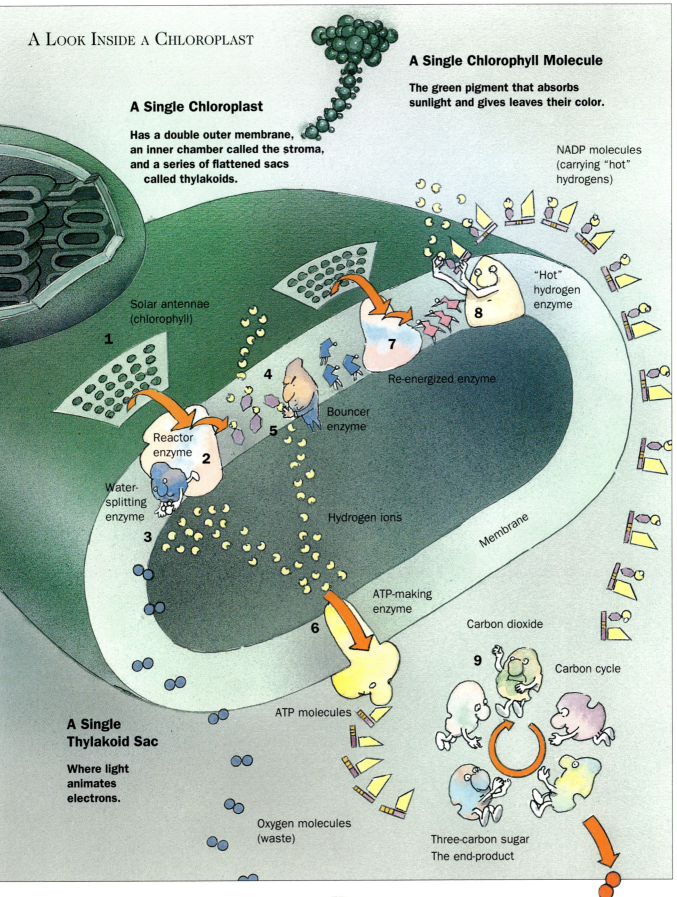

A Look Inside a Chloroplast

A Single Chloroplast

Has a double outer membrane, an inner chamber called the stroma, and a series of flattened sacs called thylakoids.

A Single Chlorophyll Molecule

The green pigment that absorbs sunlight and gives leaves their color.

NADP molecules (carrying "hot" hydrogens)

"Hot" hydrogen enzyme

8

Re-energized enzyme

7

Solar antennae (chlorophyll)

1

4

5

Bouncer enzyme

Reactor enzyme

2

Water-splitting enzyme

3

Hydrogen ions

Membrane

ATP-making enzyme

6

Carbon dioxide

9

Carbon cycle

ATP molecules

A Single Thylakoid Sac

Where light animates electrons.

Oxygen molecules (waste)

Three-carbon sugar
The end-product

Photosynthesis (continued)

Step by Step

These pages depict the steps in photosynthesis: first, in a simplified sequence (shown below); then, in a more detailed version (beginning on the facing page). We show photosynthesis as a series of steps, but you need to imagine it as a continuous, rapid flow.

The Basic Idea

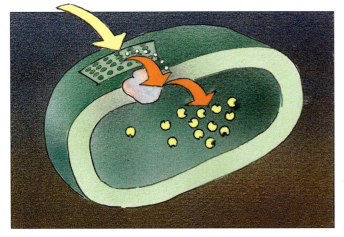

PUTTING IONS IN A SAC: Electrons, excited by sunlight, enable ions to collect in a sac.

PRODUCING ATP: The escaping ions, channeled through an enzyme, power the making of ATP.

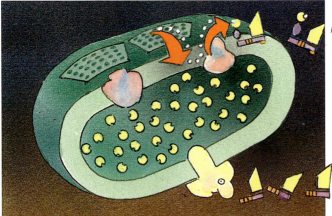

PRODUCING "HOT" HYDROGENS: The electrons, re-excited by more sunlight, are taken up by hydrogens attached to a molecule called NADP (nicotinamide adenine dinucleotide phosphate).

MAKING SUGAR: Using the energy of ATP, a circle of enzymes combine the "hot" hydrogens with carbon dioxide to make sugar.

THE DETAILS

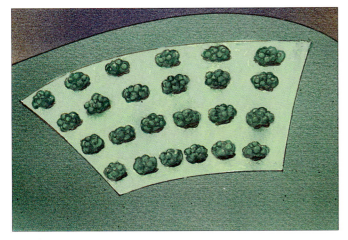

1. Chlorophyll molecules are arrayed in clusters as solar antennae.

When sunlight strikes them, their electrons jump to higher-energy orbits, bouncing around until...

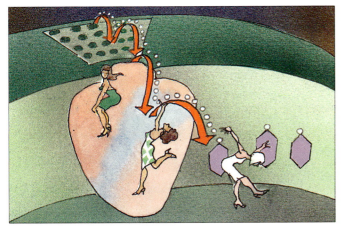

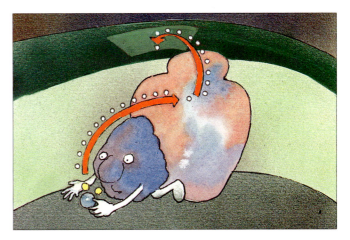

2. ...a special chlorophyll-enzyme combination transfers them to carrier molecules in the thylakoid membrane.

3. The electrons removed from chlorophyll are replaced by electrons from water...

...provided by a water-splitting enzyme that divides water into two electrons, two hydrogen ions, and one oxygen atom.

4. The electrons on the carriers attract hydrogen ions (H^+) from outside the sac, since opposites attract. Recall that a hydrogen ion plus an electron equals a hydrogen atom.

THE DETAILS

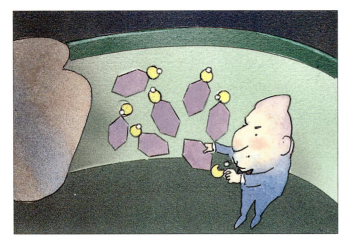

5. As the carriers reach the inner surface of the membrane, the bouncer enzyme grabs the ions...

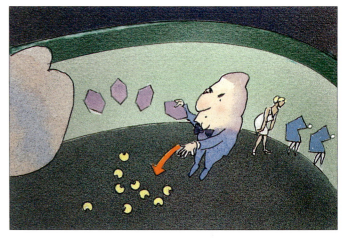

...and throws them into the sac, leaving the spent electrons to continue on.

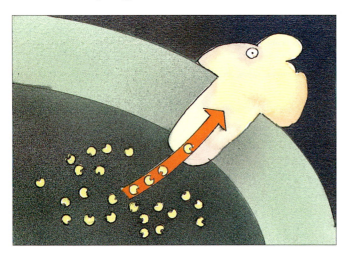

6. The only way out is via a channel through an ATP-making enzyme. The movement through this enzyme...

...supplies the energy to reattach phosphates to spent ATP molecules.

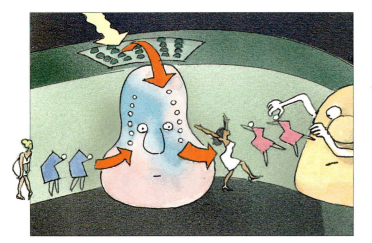

7. The spent electrons, stripped of their ions, replace electrons bouncing off a new set of chlorophylls energized by sunlight. Restored chlorophylls are ready to do more work.

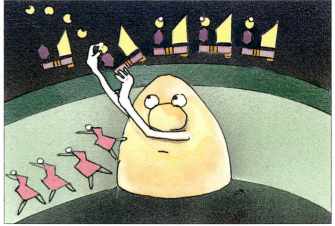

8. A final "hot" hydrogen enzyme transfers each energized electron, with a hydrogen ion, to a final carrier, NADP.

THE DETAILS

9. The action now moves to the stroma — the space in a chloroplast outside the sac. Here begins the carbon cycle, in which five enzymes cooperate to assemble half-molecules of sugar, using the ATP and the "hot" hydrogens the chloroplasts just made.

Enzyme A attaches three carbon dioxides to each of three five-carbon sugars. (The oxygens are not shown in the cycle.)

The resulting three six-carbon sugars break into six three-carbon sugars.

Enzyme B energizes the three-carbon sugar fragments with ATP.

Enzyme C attaches hydrogens to the six sugars and kicks one off the assembly line.

Enzyme D rearranges the five remaining three-carbon sugars to make three five-carbon sugars.

Enzyme E energizes these with ATP...

...and the cycle is ready to repeat.

MAKING SUGAR OUT OF THIN AIR

Perhaps the most startling thing about life is that it turns air into living substance. A team of five enzymes accomplishes this feat by transforming carbon dioxide into sugar. The enzymes each make small changes in the product molecules as they pass them around. At several key points, they use energy supplied by ATP. The "hot" hydrogens they need are on NADP. An interesting feature of this kind of enzyme-catalyzed cycle is that it always needs some of the product it's making in order to make more of that product. In this case, for every six sugar fragments that travel around one turn of the carbon cycle, only one actually rolls off the assembly line as a finished product. The other five get recycled because they are needed to initiate the first step of the cycle. A production line that makes six products only to send five of them back into the process might seem inefficient. But the enzymes work incredibly fast, producing thousands of product molecules per second.

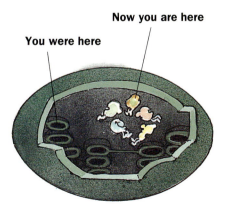

Now you are here

You were here

Making Connections

A length of pipe with only two connections (one at each end)...

...can only make a longer pipe (or backbone).

But a length of pipe with four connections...

...can make a backbone *plus* places for additional fittings. In this way, every segment of backbone can be unique.

Carbon plays so central a role in life that we say life is "carbon-based." (Twenty-four percent of the atoms in your body are carbon.) Carbon's place of honor stems from its unique ability to make four separate bonds with other atoms (i.e., share four of its electrons with other atoms). Oxygen can make only two bonds, and hydrogen one. (Nitrogen, another essential atom that comprises about 1 percent of life, can make three.) You can see the value of carbon's extra bond-making capacity through a plumbing analogy. Pipe lengths that have two connections, one at either end, can fit together to make a longer pipe but nothing else. Pipe lengths that have three or four connections can be joined to make a longer pipe with joints for additional fittings sticking out at right angles. Carbon plays a similar role in life's long-chain molecules by enabling a molecule to add both length to its backbone and connections for side groups. The backbone establishes the chain; the side groups give the chain its unique chemical character and informational value.

About Sugar

The goal of the carbon cycle is to churn out three-carbon sugar fragments (half-sugars). This marks the end of photosynthesis; but it's not the end of the line for the sugar fragments. At this point, they are shuttled through the chloroplast's outer membrane into the cell's interior, where enzymes bond them together in pairs to make the six-carbon sugar called glucose. Glucose travels through life's cellular carpet in a host of modified forms — in such disguises as sucrose, ribose, lactose, cellulose, starch, and glycogen. Glucose provides all the energy and almost all the building materials life needs.

Van Helmont's Experiment

Before the early 1600s, people assumed that the substance of plants — roots, trunk, branches, and leaves — came from the earth in which they grew. In 1630, Jean Baptista van Helmont, a Flemish physician, did a simple experiment: He planted a willow branch weighing 5 pounds in 200 pounds of soil. Five years later, after regular watering, the branch had gained 165 pounds and the soil had lost only 2 ounces! Van Helmont concluded reasonably that the material in the tree couldn't have come from the earth — that it must have come from the water. He was right in the first conclusion, but only half right in the second. It wasn't known at the time that much of the substance of life is carbon, and it didn't occur to van Helmont that the *air* might have been the source of the material in the tree. His experiment was notable, though, because by carefully weighing things, he'd at least been able to rule out earth as the source of the material. Truth comes in small pieces.

Later in life, van Helmont got interested in the gas that's produced when wood is burned. He called it "gas sylvestre" (gas from wood) but never realized it was the very carbon dioxide that nature had used to make his willow tree.

Slow Burning

Making ATP is like burning wood. When you burn real wood, you take hydrogen- and carbon-rich material, break its bonds, and combine the pieces with oxygen to produce carbon dioxide, water, and heat. When sugar is burned in mitochondria, its bonds are broken down and combined with oxygen to make carbon dioxide and water. But, in mitochondria, half of the energy is released as heat, and the other half ends up in the bonds of ATP. To accomplish this, enzymes "manhandle" the sugar to extract its hydrogen atoms. The electrons stripped from these hydrogens then flow along the mitochondrion's membranes, ultimately producing ATP.

So here's the sequence: Enzymes manipulate food (sugar fragments), extracting energetic hydrogens (1). They pass the electrons from these hydrogens along a series of carriers in a membrane inside the mitochondrion, picking up hydrogen ions as they go (2). Enzymes along the way separate the ions from the carriers and eject them into the intermembrane space (3). The accumulating ions force their way through an ATP-making enzyme (4). Finally, the spent electrons combine with hydrogen ions and oxygen to produce water (5).

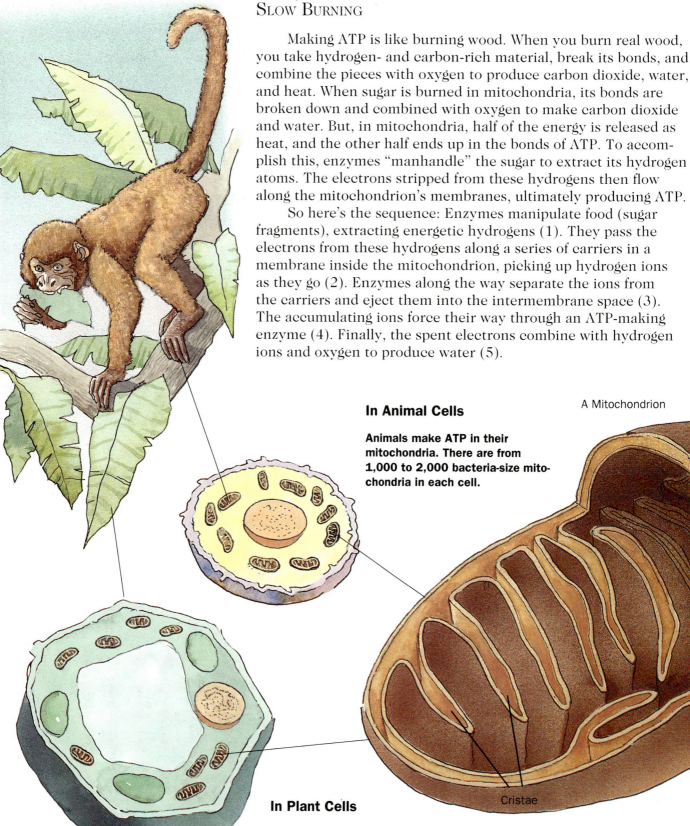

A Mitochondrion

In Animal Cells

Animals make ATP in their mitochondria. There are from 1,000 to 2,000 bacteria-size mitochondria in each cell.

Cristae

In Plant Cells

Plants produce ATP in two locations: in chloroplasts for sugar-making and in mitochondria for everything else.

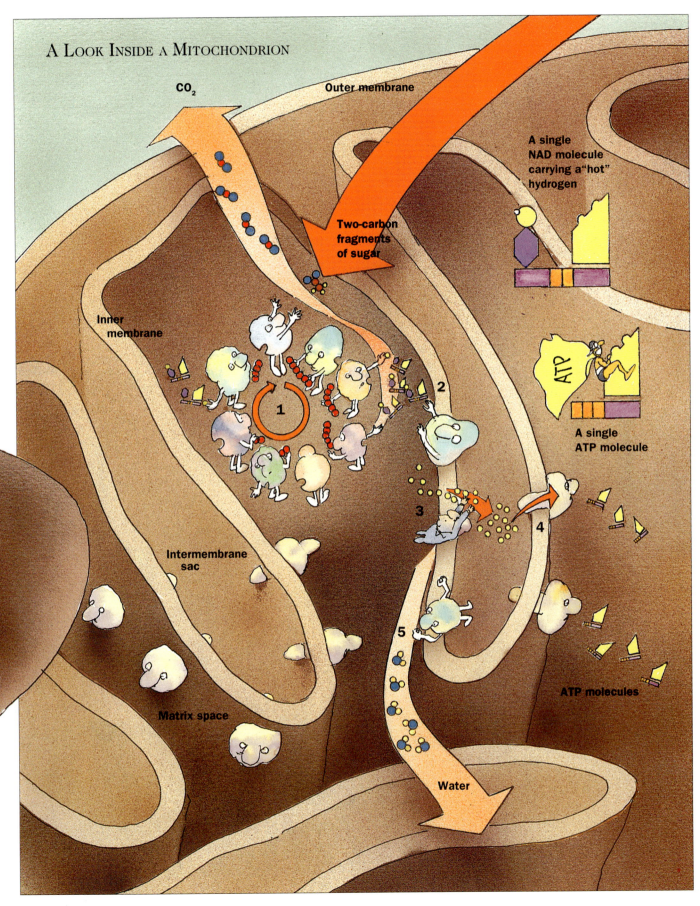

A LOOK INSIDE A MITOCHONDRION

CO$_2$

Outer membrane

A single NAD molecule carrying a "hot" hydrogen

Two-carbon fragments of sugar

Inner membrane

ATP

A single ATP molecule

1

2

3

4

5

Intermembrane sac

ATP molecules

Matrix space

Water

RESPIRATION (CONTINUED)

STEP BY STEP

Respiration in a mitochondrion is similar to photosynthesis in a chloroplast — but almost in reverse. Both processes involve a cycle of enzymes manipulating molecules and a flow of electrons in a membrane. The numbers on the map at the right correspond to the numbered locations on pages 67 and 68.

THE BASIC IDEA

REMOVING "HOT" HYDROGENS: Enzymes remove hydrogens from sugar, putting them on carriers.

STARTING AN ELECTRON FLOW: Another enzyme strips electrons from the "hot" hydrogens.

PUTTING IONS IN A SAC: The flowing electrons attract hydrogen ions, which get pushed into the intermembrane sac in the cristae.

PRODUCING ATP: The force of the escaping ions channeling through an enzyme produces ATP.

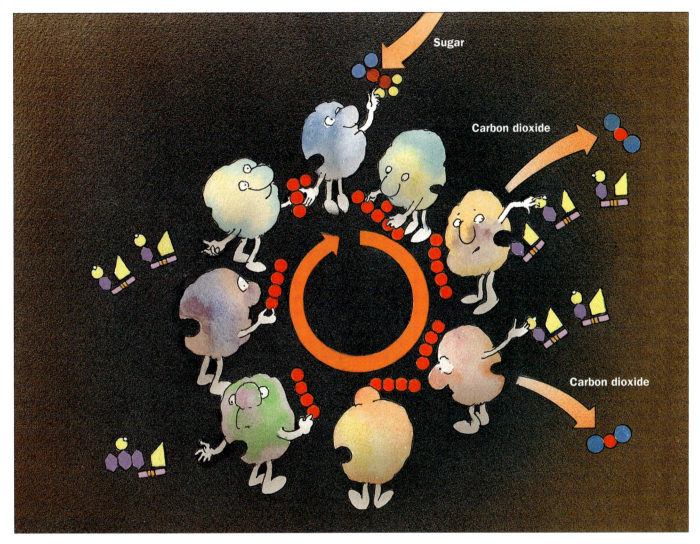

1. Sugar enters the mitochondrion as two-carbon fragments (top), having been broken off from glucose (six carbons) in a process called glycolysis (see page 70). Several enzymes then manipulate these fragments to extract "hot" hydrogens at several points in the cycle. Carbons and oxygen combine and are discarded as carbon dioxide — which animals exhale.

2. "Hot" hydrogens, passing in an endless stream, give up their electrons to an enzyme in the inner membrane...

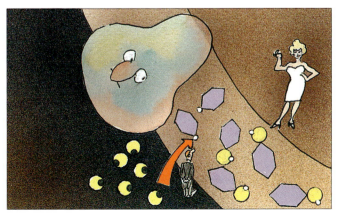

...which passes the electrons to carrier molecules floating in the membrane. Each electron is picked up by a hydrogen ion (making it a hydrogen atom on the carrier).

The Details

3. The carriers randomly "dance" the energized hydrogens over to the next enzyme...

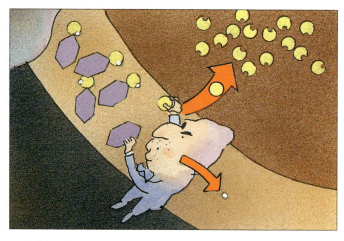

...which takes the electrons off the carriers and passes the ions into the intermembrane sac.

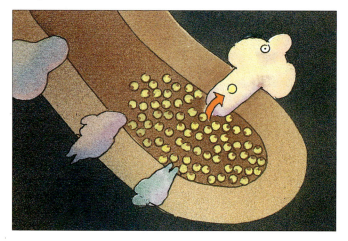

4. As the ions accumulate in the intermembrane sac, some of them escape through ATP-making enzymes.

The ions, moving through the enzyme, energize it, and it can then reattach phosphates to spent ATP molecules.

So a steady stream of rejuvenated ATP molecules emerges, ready to supply the energy for the work of the cell.

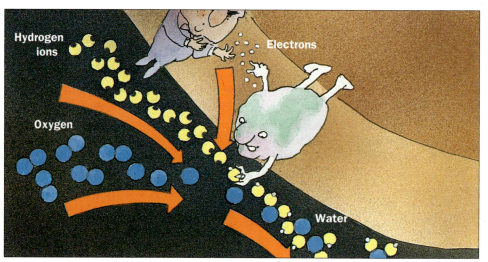

5. Finally, a third enzyme combines the spent hydrogen ions and oxygens to form water — a waste product of the process.

Oxygen Discovered

Since earliest times, flames arising from burning material were seen as evidence that something essential was being *released*. This something became known as "phlogiston" in the 1700s. Scientists found that if they burned material inside an enclosed space, the flames would soon gutter out. What's more, the air inside the enclosure could no longer sustain an animal's life. It appeared that accumulated phlogiston inhibited both fire and life. Scientists found they could rejuvenate this "phlogisticated" air by putting plants in it — if the plants were exposed to sunlight. The plants somehow counteracted the effects of phlogiston.

The great French chemist Antoine Lavoisier decided to investigate the nature of phlogiston. In the 1780s, he carefully measured the amounts of everything involved in burning a given material. He showed, for instance, that when he set fire to a piece of metal, it melted but actually *gained* weight; moveover, this increase in weight was *exactly equal* to the loss in weight of the surrounding air. (When wood is burned, the ash is, of course, *lighter* than the original log. This is because, unlike metal, wood's cellulose combines with oxygen to form carbon dioxide and water, which escape as part of the smoke. Add up the weights of the ash and the escaped gases, and they, too, would exceed the weight of the wood.)

Lavoisier later named the gas in air that combined with the metal: oxygen. It all became clear. Air, depleted of oxygen by fire, can no longer sustain fire or life because these processes require oxygen. The photosynthesizing plant rejuvenates the air by excreting oxygen into it.

LET'S HEAR IT FOR OXYGEN

When oxygen first appeared in the earth's atmosphere, it was poisonous to most organisms because it produced ionized molecules that could damage DNA. But, as often happens in evolution, adverse conditions create new opportunities. The "respirers" — those organisms that adapted to oxygen by evolving various ways to neutralize the unwelcome ions — thrived spectacularly.

Oxygen greatly expands an organism's ability to produce energy. A "fermenter" — a microorganism that doesn't use oxygen (see next page) — can get only two ATPs out of one glucose molecule. A respirer can get twenty! With this advantage, the respirers, which include many bacteria and virtually all multicellular organisms, have come to dominate the living world.

There's one odd thing about oxygen's role in respiration: It is *hydrogen* — actually its electrons and ions — not oxygen, that an organism needs to produce energy. But the organism must have a way to dispose of its steady stream of spent hydrogens. That's where oxygen comes in, combining with the leftover hydrogen to make water (H_2O). So oxygen, on which our lives so totally depend, is not really in the show; it arrives at the stage door when the play's over, to pick up the exhausted performers.

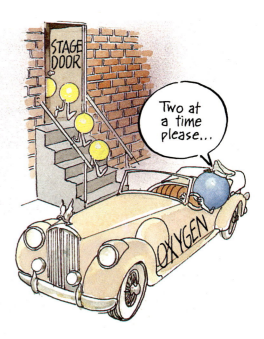

GLYCOLYSIS

ENERGY WITHOUT OXYGEN

In the earth's ancient oceans, before anything could perform photosynthesis, organisms developed ways of getting energy from sugar *without the need for oxygen*. This operation was a primitive form of what we recognize today as glycolysis in animal cells or fermentation in microorganisms. In these processes, each molecule of glucose gets broken into smaller pieces by a series of enzymes, generating two ATP molecules. While this amount of ATP is much less than the twenty molecules that animal cells can make in their mitochondria by further breaking down the fragments of a molecule of glucose, it is an important emergency energy source for animals. For instance, when there's a sudden demand for heavy muscular work (as in running the 100-yard dash) and there's not enough time for oxygen to be delivered to the muscle cells by the blood, glycolysis supplies the necessary ATP.

A primitive form of glycolysis, consuming plentiful sugar-like materials in the ancient oceans before photosynthesis and oxygen became available, was probably life's earliest way of producing usable energy.

In the Plant Cell ▶

The half-molecules of glucose made by chloroplasts are paired to make glucose in the cytoplasm, and stored in other forms such as sucrose and starch. The glucose molecules are broken down as needed into two-carbon pieces by glycolysis. The pieces are further broken down in plant mitochondria to make ATP, which supplies the energy for the plant's activities.

munch munch munch

In the Animal Cell ▶

Animals eat plants and the liberated glucose molecules, broken down into two-carbon pieces by glycolysis, are further broken down in animal mitochondria to make ATP, which supplies the energy for the animal's activities.

Pasteur's Wine

Interest in fermentation — the breakdown of sugar to alcohol — goes far back in human history. Until 1860, it was believed to be a purely chemical process, having nothing to do with life. Then Louis Pasteur showed that fermentation was a living process carried out by yeast and bacteria. He found he could prevent spoilage of wine and beer by pasteurization — the process of using heat to destroy bacteria that produced vinegary, unpalatable acids. His achievements not only benefited the French wine and beer industries but led to discoveries in the early 1900s that showed that fermentation also occurred in animals and plants. The general process was called glycolysis, meaning sugar breakdown, and was found to be accompanied by the formation of ATP. Soon it was shown that contraction of muscle tissue goes hand-in-hand with glycolysis and ATP production, and that ATP is quickly consumed as the muscle continues to work. These discoveries led to our present understanding: Glycolysis, which needs no oxygen, and respiration, which needs oxygen, both synthesize ATP, which supplies the energy for all cell work.

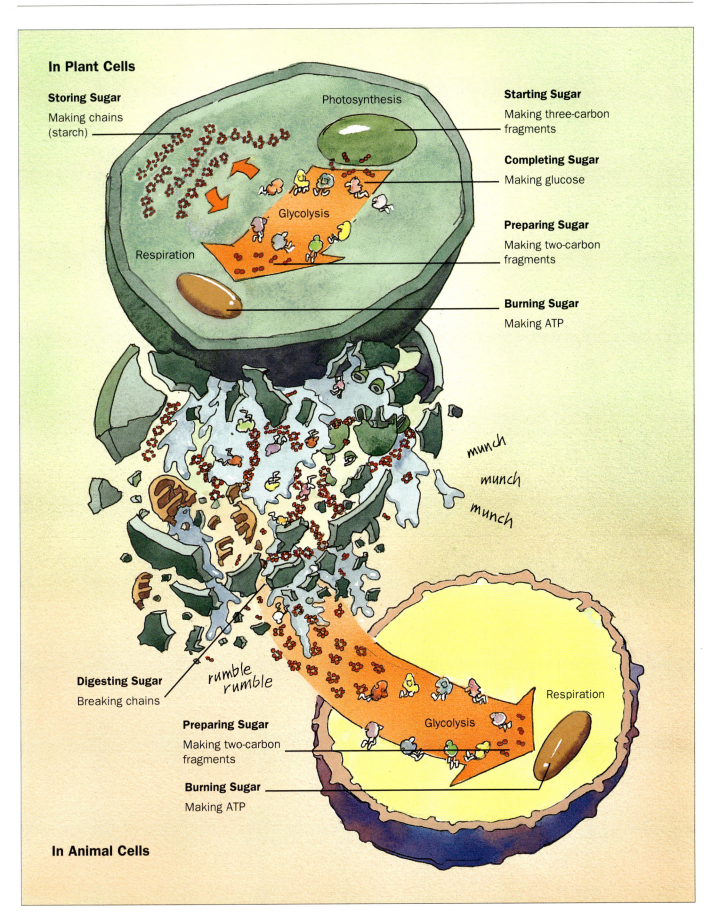

In Plant Cells

Storing Sugar

Making chains (starch)

Photosynthesis

Starting Sugar

Making three-carbon fragments

Completing Sugar

Making glucose

Glycolysis

Preparing Sugar

Making two-carbon fragments

Respiration

Burning Sugar

Making ATP

munch

munch

munch

rumble rumble

Digesting Sugar

Breaking chains

Preparing Sugar

Making two-carbon fragments

Respiration

Glycolysis

Burning Sugar

Making ATP

In Animal Cells

COMMUNITY ENERGY

LIVING LIGHT

Over the past 4 billion years, cells that were once free-living have joined together in cooperating communities that we call multicellular organisms. Within these organisms, groups of cells have taken on special roles — becoming muscle, brain, bone, skin, etc. These extra duties often require the additional production and consumption of energy as ATP. Instead of spending energy as lone hunter-gatherers, specialized cells divert some of their ATP to the performance of "civic" duties that benefit the whole community. Individual cells in a firefly's tail, for example, have no need to light up; since they are members of a larger community that must mate and reproduce, however, their light-making becomes a vital, ATP-requiring activity.

From a Cellular Perspective

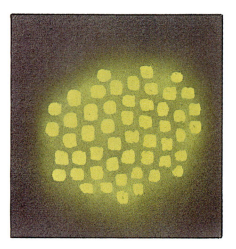

From a Molecular Perspective

In the cells of the firefly tail, an enzyme attaches a part of ATP to a molecule called luciferin. This energizes luciferin, allowing oxygen (O_2) to bond to one of its carbon atoms and boosting an electron into a higher orbit. Luciferin then releases the oxygen and carbon as carbon dioxide (CO_2). As the electron drops back to its customary orbit, the spent energy is released as a tiny flash of light. This process, which uses ATP and oxygen to produce light and carbon dioxide, is the exact opposite of photosynthesis!

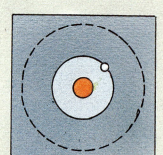

Electron in normal orbit

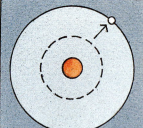

Electron boosted to a higher orbit

Electron falling back to a lower orbit, releasing energy as light

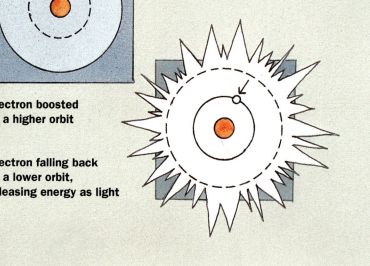

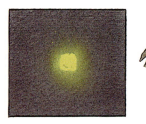

Every cell in a firefly must produce ATP for its own needs.

In addition, the cells in the firefly's tail must make extra ATP so that the whole tail can glow.

When millions of tail cells glow, the firefly has a good chance of mating and producing more fireflies.

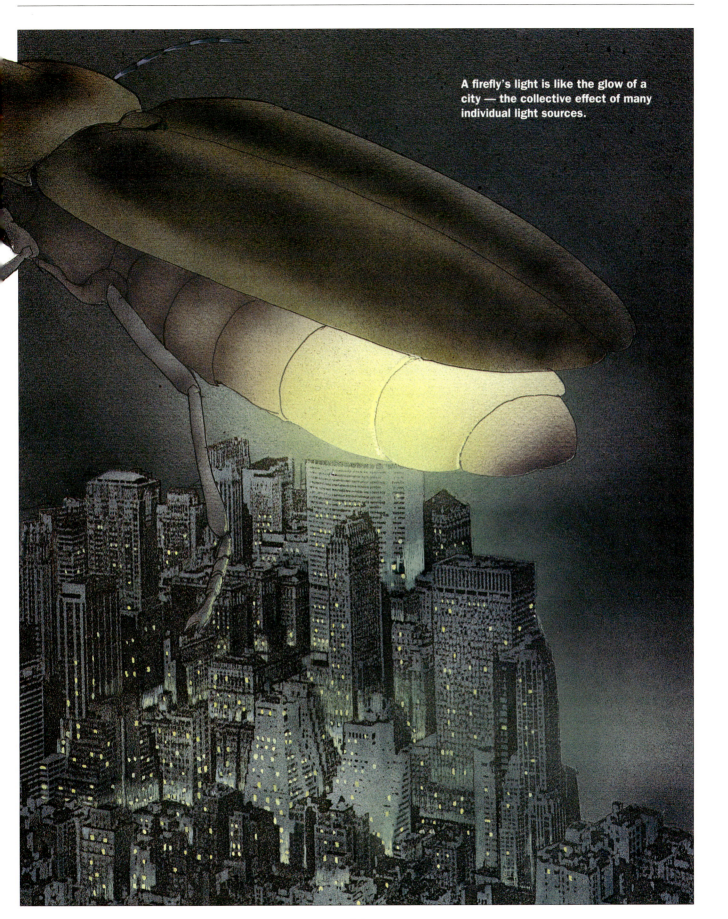

A firefly's light is like the glow of a city — the collective effect of many individual light sources.

73

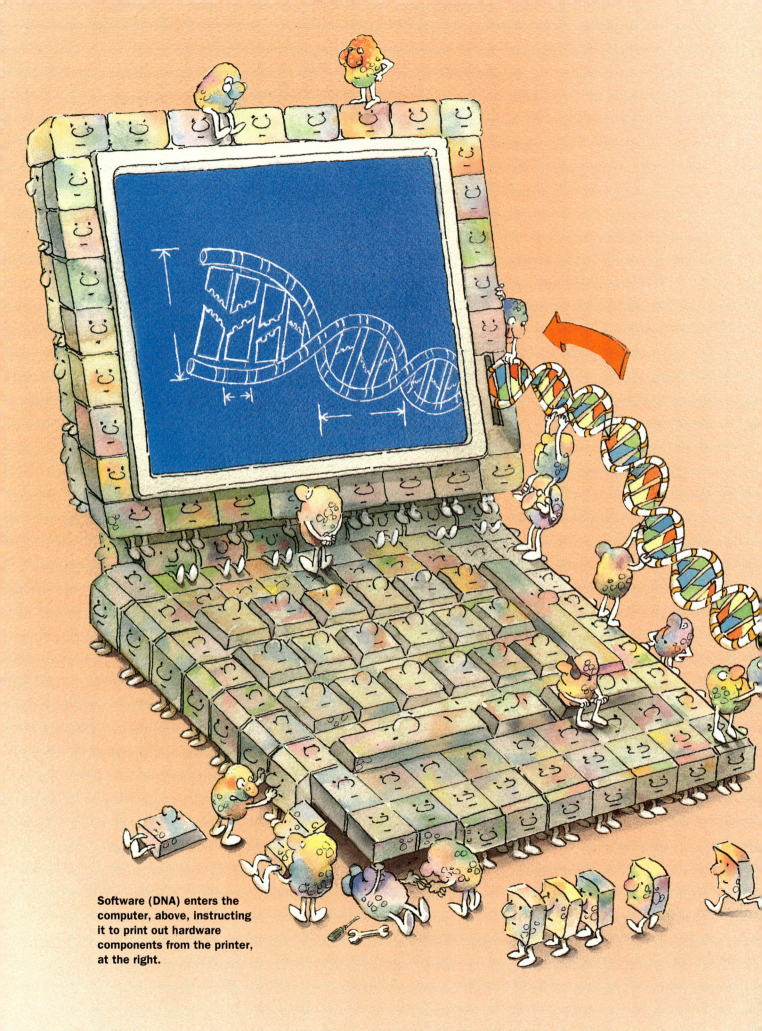

Software (DNA) enters the computer, above, instructing it to print out hardware components from the printer, at the right.

Chapter 3

INFORMATION

The Storehouse of Know-How

Imagine a computer that can build itself. Its software — the programs that direct its machinery, or hardware — contains the recipe for its own assembly. After it finishes making all of its own hardware from the software's instructions, it maintains and repairs itself — until it begins to perceive the end of its workable lifetime. Then it makes a copy of its carefully stored software, which, in turn, proceeds to make a replica of itself. And the cycle repeats itself indefinitely.

A living cell is such a self-organizing system. As such, it embodies a paradox: If hardware depends on software, and software depends on hardware, how could such a process ever get started in the first place? You'll recognize this as another version of the classic question, "Which came first, the chicken or the egg?" (see page 98).

More puzzles center on the nature of the software that life uses. For example, where does a young sapling store the information needed to build another oak tree? And how does this information interact with the rest of the tree's hardware? In this and the next chapter, we explore the relationship between software and hardware, between the "ideas" and the "working parts" of life. We use the terms "information" and "machinery."

It is no more likely that life could arise from non-life — for example, that flies could be created by decaying meat — than that a 747 could accidentally be assembled by a tornado blowing through a junkyard.

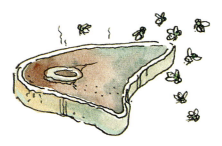

Flies, of course, pose an assembly problem much tougher than a 747. They are the result of billions of years' worth of accumulated "research and development," i.e., information built up by trial and error. Flies, like airplanes, can't be built without lengthy assembly instructions.

AN UNBROKEN CHAIN

Throughout most of history, people believed life was controlled by mysterious, supernatural forces. For instance, seeing worms, maggots, flies, or even mice squirming around in decaying grains, mud, or rotting meat convinced early scientists that life arose spontaneously from non-life. Carefully controlled experiments later refuted this notion (see the opposite page), but it still took a long time to appreciate why, even in principle, "spontaneous generation" is impossible: Life must always come from life, flowing from generation to generation in an unbroken chain. This inevitable conclusion comes from our modern understanding of the key role played by the information life stores within itself.

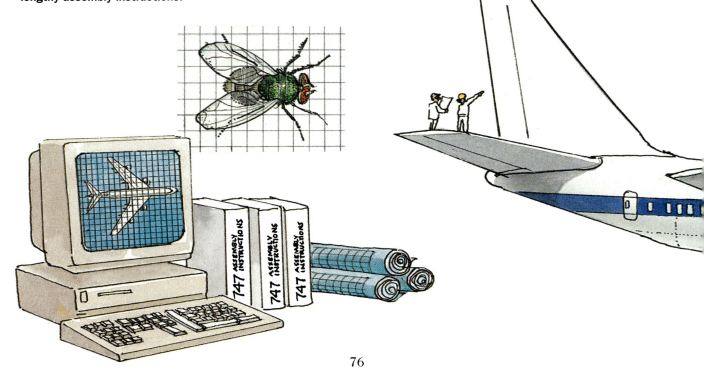

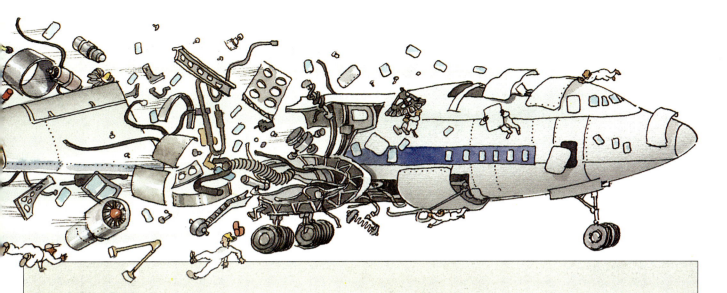

The Death of "Spontaneous Generation"

In 1668, in one of the earliest carefully controlled biological experiments, Francesco Redi took eight flasks with meat in them, left four open to the air, and sealed the others. After a while, maggots infested the four open flasks, but none appeared in the sealed flasks. Redi then tried covering the formerly sealed flasks with gauze: no maggots. He correctly concluded that the maggots had come from eggs deposited on the meat by flies. This experiment disproved the notion that visible creatures, at least, were spontaneously generated out of nothing in decaying matter.

Yet people still believed that microorganisms such as bacteria and yeast sprang spontaneously from decaying matter. The controversy raged on until 1864, when Louis Pasteur determined to end it. First, he tested air and dust and showed that they contained living organisms. Next, he added air and dust to thoroughly sterilized materials, sealed them in flasks, and noted that life rapidly multiplied within. Then he placed sterilized matter inside a flask with a long S-shaped neck stopped up with a cotton plug; no life arose inside the flask. If Pasteur tipped the flask so that its contents touched the cotton plug, thereby contaminating them with microorganisms trapped in the plug, life sprang forth inside the flask within 48 hours. And, of course, if he broke the neck and let air in, growth inside the flask rapidly ensued. Said Pasteur, "Never will the doctrine of spontaneous generation recover from the mortal blow that this simple experiment has dealt it." It didn't.

Uncovering the Secrets of Heredity

Once scientists realized that life can only come from life, they began to look more closely at inheritance. Yes, our offspring look like us…but why?

This short survey takes us up to the 1940s.

1860s

"Factors" Determine Inheritance

Austrian monk Gregor Mendel discovers that something he dubs "factors" somehow determine inheritance in pea plants. Every trait appears to be controlled by a pair of these factors. Further, a trait may have "dominant" and "recessive" forms. For instance, if Mendel bred a tall plant with a short one, the offspring were mostly tall; tallness is dominant, and shortness recessive. However, the recessive trait isn't lost — it can reappear in a later generation; two tall pea plants bred together might produce a short one.

1890s

Chromosomes

Chromosomes, microscopic structures in the cell nucleus, are discovered by many researchers. They note that chromosomes, which come in pairs, double before cell division and are then shared between daughter cells. It is suspected that chromosomes are the carriers of heredity.

1903

"Factors" Are on Chromosomes

William Sutton makes the connection between Mendel's factors and chromosomes. One member of each pair of trait-determining factors is on one of a pair of chromosomes. One chromosome comes from the mother's egg, and the other from the father's sperm.

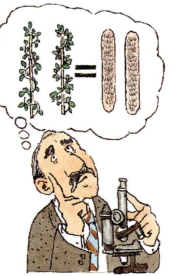

1905

Chromosomes Actually Determine Inheritance

Edmund B. Wilson and N. M. Stevens discover that a particular chromosome called the X chromosome, of which there are two in female cells and one in male cells, determines the sex of the offspring and explains why there are equal numbers of females and males: All eggs get an X, but only half of sperm do (the other half get a Y chromosome). This is the first evidence that a specific chromosome carries a specific hereditary property (sex).

1906

Mendel's "Factors" Are Genes

Scientists coin the term "gene," meaning a piece of genetic information specifying a particular trait or characteristic. Genes are the factors Mendel discovered.

Genes Are Inherited Together

Thomas Hunt Morgan shows that many genes are inherited together, as would be expected if they are linked to each other in chromosomes. (The fruitfly has four chromosomes, and it has four groups of linked genes.) Chromosomes, then, are chains of genes.

1908

Genes Are Lined Up Along Chromosomes

Morgan observes that even though genes tend to be inherited together, this occurs more frequently with some pairs than with others. He infers that the farther apart genes are on a chromosome, the less likely they are to be inherited together. (This is because an actual physical exchange of genes takes place between chromosomes; see page 201.) Morgan is able to "map" the relative positions of genes along fruitfly chromosomes.

1909

Hereditary Diseases May Be Caused by Defective Genes

Archibald Garrod postulates that certain inheritable human diseases result when particular proteins fail to perform their normal function.

New Traits Are Caused by Mutations

Scientists realize that mutations — changes in genes — are what produce new genetic characteristics (as well as inherited diseases). They further realize that without mutations, there can be no evolution (see page 202). Hugo de Vries had discovered genetic mutations in 1886. Hermann Muller first produces mutations with x-rays in 1927.

1927

One Gene — One Protein

George Beadle and Edward Tatum, using bread molds, show that individual genes control production of individual proteins (see page 83).

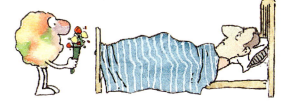

1942

1944

Natural Selection Operates on All Living Things

Salvador Luria proves that bacteria are subject to the same genetic and evolutionary forces that operate on plants and animals. Bacteria, because they reproduce so rapidly, become the main experimental subject of molecular genetics (see page 216).

Genes Are Made of DNA

Oswald Avery and associates show that genes are made of deoxyribonucleic acid — DNA.

THE MAP AND THE TERRITORY

Here we've listed some instances of the relationship between encoded "ideas" and their decoded "products."

Idea	Product
Map	Territory
Blueprint	Building
Recipe	Cake
Menu	Meal
Sheet Music	Symphony
Genes	Proteins

We can consider the items in the left-hand column as chunks of information corresponding to the actual products in the right-hand column. But, strictly speaking, coded information also exists in material form — ink, paper, molecules — so the items on the left are also products.

He's just full of information.

Yep...or grasshoppers.

Information Is Embedded in Living Things

A fundamental difference between living and non-living things is that living things use information to create and maintain themselves. Rocks contain no instructions on how to be rocks. Toads contain instructions on how to be toads.

Information, like ideas, is dimensionless. It's simply a comparison between one thing and another, a registering of differences. Information becomes tangible when it is encoded in sequences of symbols: zeros and ones, dots and dashes, letters of the alphabet, musical notes, etc. Such sequences of symbols, in turn, are decoded — by machinery or by us — into computer output, Morse code messages, books, symphonies, etc. In order to be stored or transmitted, then, information needs to be put into some physical form, a process that requires energy. In this sense, you might say that "mind" and "matter" are inextricably linked.

Life's information — the "ideas" governing how it operates — is encoded in genes, which are, in turn, decoded by machinery that manufactures parts that work together to make a living creature. Like the computer that builds itself, the process follows a loop: Information needs machinery, which needs information. This relationship can start simply and then, over many generations, build into something complex. Similarly, our deeper thoughts evolve out of simpler bits and pieces — hunches, ideas, memories.

Rocks are simple, stable arrangements of molecules settled into low-energy states.

A toad's cells are complex arrangements of high-energy molecules, dynamically organized by information.

Information Needs Difference

A chain that simply repeats one symbol carries no useful information.

This tells me nuthin'!

But a chain made up of different symbols can encode information. All of life's genetic instructions are spelled out in combinations of four different "letters."

Interesting...

DNA — What Does It Actually Say?

Not a Blueprint But a Recipe

We might never understand life's complexity were it not for the discovery that life is orchestrated by "intelligent" worker molecules called proteins. These proteins are various combinations of twenty, and only twenty, different amino acid molecules linked together into long chains. Every unique function of a protein is determined by the order of the amino acids in its chain.

Here we have a powerful insight into the way life works: One chain, DNA, carries information; a second chain, of amino acids linked into proteins, does life's work of growing, maintaining itself, and reproducing. DNA's sequence of units determines the sequence of amino acid units in proteins. Thus, DNA is not like a blueprint, which contains an image or a scale model of the final product; it is more like a recipe — a set of instructions to be followed in a particular order.

So life's complexity arises from a breathtaking simplicity: DNA's message says, "Take this, add this, then add this…stop here. Take this, add this, then add this,…etc." While the idea is simple, accomplishing it requires some ingenious machinery (see Chapter 4).

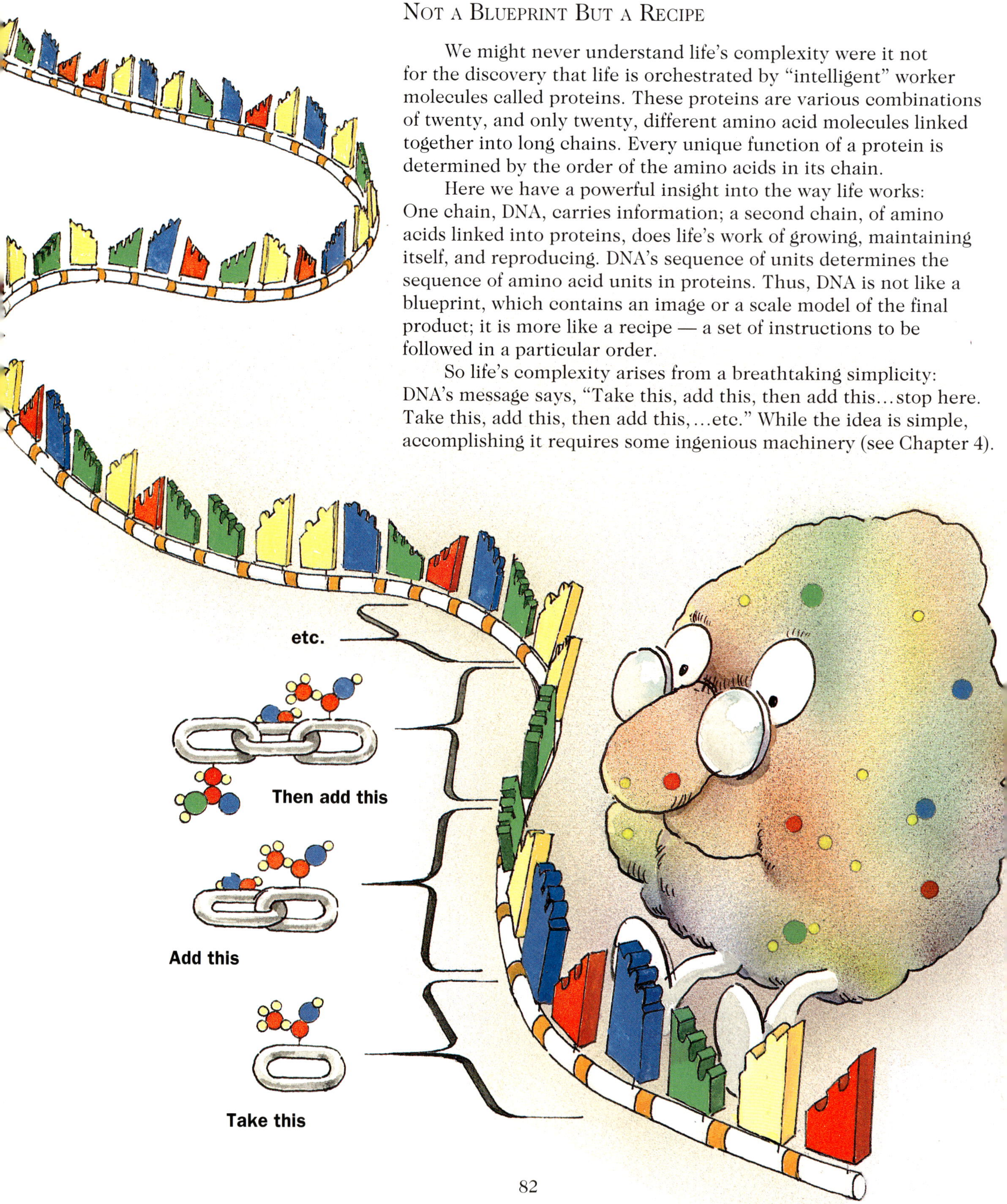

etc.

Then add this

Add this

Take this

82

A Key Discovery: One Gene Makes One Protein —
Beadle and Tatum and Bread Mold

Most of the inherited traits studied up to the early 1940s were complex functions: height of pea plants, fruitflies' wing shape or eye color, etc. These were probably controlled by many genes.

George Beadle realized he had to narrow the focus — to find one simple trait controlled by one specific gene. Inspired by Thomas Morgan, he started working with fruitflies but soon found a better subject — the common bread mold Neurospora. Here's the kind of experiment he and his associate, Ed Tatum, did. Normal molds can convert sugar, step-by-step, into all twenty amino acids. For instance, amino acid X is made by converting molecule A into molecule B, then B into C, and finally C into X. Beadle and Tatum exposed molds to x-rays, causing them to change — to mutate — so they could no longer make certain amino acids. One mutant could no longer make X unless Beadle and Tatum supplied it with molecule C. Giving it A or B didn't help. They concluded that the mutant had lost the ability to convert B to C — in other words, x-rays had damaged the enzyme that converts B to C. Another mutant was unable to make X unless it was supplied with molecule B. Beadle and Tatum concluded that this mutant had lost its ability to convert A to B — that is, x-rays had damaged the enzyme that converts A to B. Beadle and Tatum correctly surmised that each mutant had sustained x-ray damage to a specific gene that was responsible for making a specific enzyme. This simple idea — that one gene codes for one enzyme — opened the door to deeper understanding of how genes work.

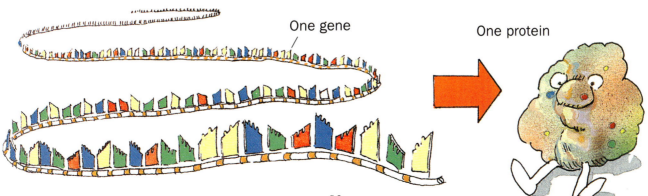

One gene

One protein

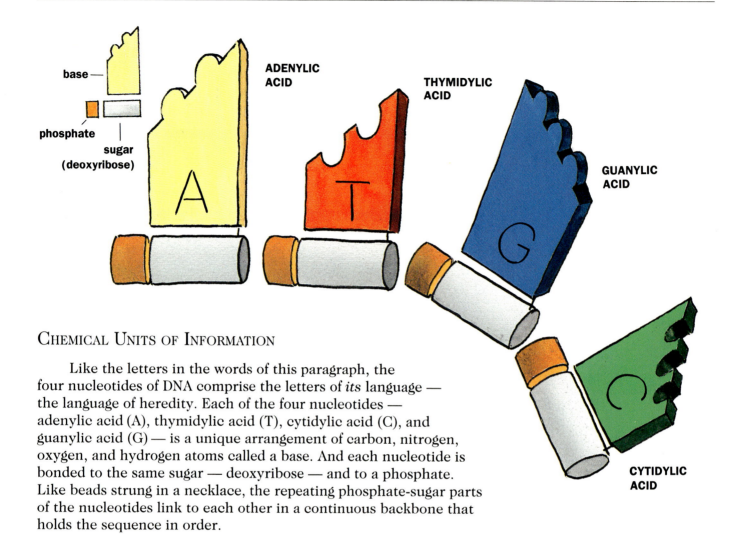

base

phosphate

sugar
(deoxyribose)

ADENYLIC ACID

THYMIDYLIC ACID

GUANYLIC ACID

CYTIDYLIC ACID

CHEMICAL UNITS OF INFORMATION

Like the letters in the words of this paragraph, the four nucleotides of DNA comprise the letters of *its* language — the language of heredity. Each of the four nucleotides — adenylic acid (A), thymidylic acid (T), cytidylic acid (C), and guanylic acid (G) — is a unique arrangement of carbon, nitrogen, oxygen, and hydrogen atoms called a base. And each nucleotide is bonded to the same sugar — deoxyribose — and to a phosphate. Like beads strung in a necklace, the repeating phosphate-sugar parts of the nucleotides link to each other in a continuous backbone that holds the sequence in order.

From Nucleotide to Genome — A Hierarchy of Packaging

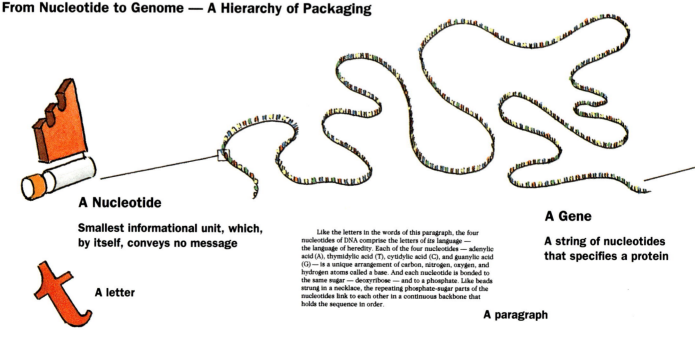

A Nucleotide

Smallest informational unit, which, by itself, conveys no message

A letter

Like the letters in the words of this paragraph, the four nucleotides of DNA comprise the letters of *its* language — the language of heredity. Each of the four nucleotides — adenylic acid (A), thymidylic acid (T), cytidylic acid (C), and guanylic acid (G) — is a unique arrangement of carbon, nitrogen, oxygen, and hydrogen atoms called a base. And each nucleotide is bonded to the same sugar — deoxyribose — and to a phosphate. Like beads strung in a necklace, the repeating phosphate-sugar parts of the nucleotides link to each other in a continuous backbone that holds the sequence in order.

A paragraph

A Gene

A string of nucleotides that specifies a protein

A Chemical Can Genetically Change Cells

In 1928, Frederick Griffith, a London medical officer, made a momentous discovery. At that time, the major cause of death worldwide was lobar pneumonia, caused by the pneumococcus bacterium. Scientists knew that certain mutant forms of these bacteria were benign; i.e., they didn't cause disease. Griffith discovered that if he mixed these *living* harmless pneumococci with dead disease-causing pneumococci and injected the mixture into mice, the mice all died of pneumonia. Moreover, their bodies were teeming with living, multiplying killer pneumococci! Something had been released from the dead killer cells and got inside the living benign cells and changed their inheritance; harmless cells had been permanently *transformed* into killer cells. What was this transforming substance? Griffith was never to know — he died in the bombing of London in 1941.

It took many years of laborious chemical analysis and the painstaking development of methods for purifying and testing cell components before Oswald Avery, Colin MacLeod, and Maclyn McCarty at the Rockefeller Institute in New York announced, in 1944, that the transforming agent was DNA. Their work confirmed that DNA is the genetic molecule; genes are made of DNA.

Dead killer pneumococci + **Live harmless pneumococci** = **Live killer pneumococci**

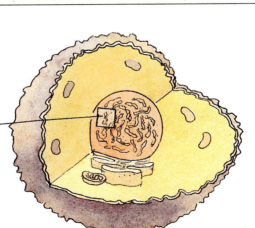

A Chromosome

A spooled-up string of genes (about 3,000) packaged in a single unit

One volume

A Genome

All of the chromosomes of a single organism — usually collected in the nucleus of each of its cells

A set of volumes

DNA — BASE PAIRS AND WEAK BONDS

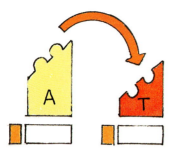

The nucleotides adenylic acid and thymidylic acid...

...go together in a perfect fit.

NUCLEOTIDE PAIRS —
A KEY TO STRUCTURE AND FUNCTION

DNA is always found as a double chain of one sequence of nucleotides paired with another sequence of nucleotides. As you can see, the base parts of the four nucleotides (marked A, T, G, and C) match up in pairs. Their shapes and chemical make-up are such that A fits only with T and G fits only with C. These pairs, when fitted together, have exactly the same width (the distance from sugar to sugar). So the sequence of nucleotides in one chain of DNA will exactly match a complementary sequence in the other chain — and the two chains will always be exactly the same distance apart. For example, if the sequence of nucleotides on one side is G-T-A-C-C, the sequence on the other side is C-A-T-G-G.

Moving In on the Structure of DNA

James Watson and Francis Crick, who began to work together in Cambridge, England, in 1951, believed that if they could visualize the form of a DNA molecule, they might see how it carried information and how it made copies of itself. They already knew a lot about DNA's chemistry. DNA was first discovered by Johann Miescher in Switzerland back in 1869, and, over the years, many chemists had identified its four nucleotides and found out how they were linked in a chain. Furthermore, in 1949, Irwin Chargaff, a chemist at Columbia University in New York, had shown that while samples of DNA taken from different organisms — animals, plants, yeast, or bacteria — contained different amounts of the four nucleotides, the amount of adenylic acid in each sample *always* equaled the amount of thymidylic acid, and the amount of guanylic acid *always* equaled the amount of cytidylic acid. At the time, no one knew why the quantities of these pairs of nucleotides showed this consistent relationship. What structure could account for this property?

The nucleotides guanylic acid and cytidylic acid...

...also go together in a perfect fit.

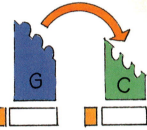

One chain is a counterpart, or complement, of the other. Note, too, that because of the way A and T or G and C must match up, the two chains must have opposite chemical directions — indicated at the right by the opposing arrows.

Nucleotide pairing enables the two chains of DNA to fit together perfectly.

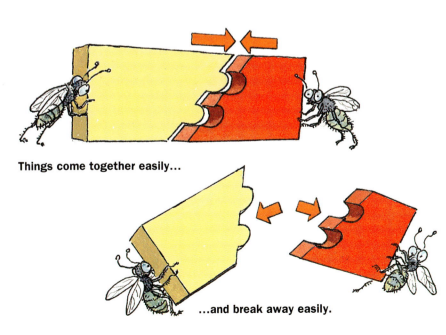

Things come together easily...

...and break away easily.

WEAK BONDS

Weak bonds make it possible for big molecules to change shape or come apart and rejoin. Twenty times weaker than the covalent bonds that hold atoms together in molecules, these weak attractions between positive and negative charges can form only at very close range. Such weak bonds hold A to T and G to C in DNA and allow the two chains to separate readily, which they must do to replicate themselves.

DNA — The Double Helix

Information with a Twist

DNA resembles a spindly ladder that has been twisted so that its sides form spirals. Its exceeding thinness means it can easily be packed into small places. Its doubleness ensures that it won't get tangled up in itself; and it also protects the precious inward-facing nucleotide sequence — DNA's language — from damage. And, as we shall see, this doubleness is essential for DNA to be copied.

Bacteria carry their DNA in one long double helix. In our cells, the DNA resides in 46 chromosomes — 46 double-stranded helices. The chains are stupefyingly long: If we think of the links in each DNA chain as letters, bacterial DNA represents about 40 average novels; human DNA about 40,000! If all of the DNA in *one* of our cells was laid out end to end, it would be about 2 yards long. Since we have about 5 trillion cells, the *total* length of DNA in each of us would reach the 93 million miles from here to the sun 60 times. For a double chain that long to fit into a space as small as a cell nucleus, it must be incredibly thin.

The molecular model at the left shows DNA's constituent atoms.

Watson and Crick Discover the Structure of DNA

In London in 1952, Maurice Wilkins and Rosalind Franklin were using a process called x-ray diffraction to examine the shape of DNA. They shone x-ray beams through DNA and recorded on photographic film the pattern of scattering caused by the DNA molecules. Their work suggested that DNA was in the form of two or three chains whose bases somehow stacked near one another.

At Cambridge, Watson and Crick made cardboard and then sheet metal cut-outs of the nucleotides, based in part on knowledge obtained by Rosalind Franklin and Maurice Wilkins. This model-building approach was a key to their ultimate success.

A big eye-opener came when Watson and Crick learned that the *molecular* shapes of DNA's nucleotides were such that adenylic acid fit *only* with thymidylic acid and guanylic acid fit *only* with cytidylic acid. This made sense of Chargaff's discoveries. When Watson and Crick "mated" these base pairs inside DNA's sugar-phosphate backbones in a double helix, everything fit beautifully.

Watson and Crick triumphantly presented their model to the scientific world in 1953. Its acceptance was immediate, not only because of its intrinsic elegance but because it at once suggested how DNA could replicate itself: One strand was a complementary copy of the other; if the two strands were separated, new nucleotides could be laid down along each to form new strands (see the next page).

DNA — Creating Its Own Future

Doubling of Information

Before a cell divides to become two, its DNA must be doubled so that each daughter cell will receive a perfect copy. This means the strands of DNA must first be separated, then complementary nucleotides must be linked along each of the separated strands.

DNA Replication — The Basic Idea

A double strand of DNA...

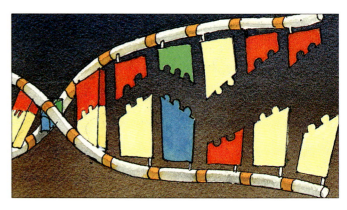

..."unzips" like a zipper.

Free-floating nucleotides match with their complements...

...and connect along the backbone.

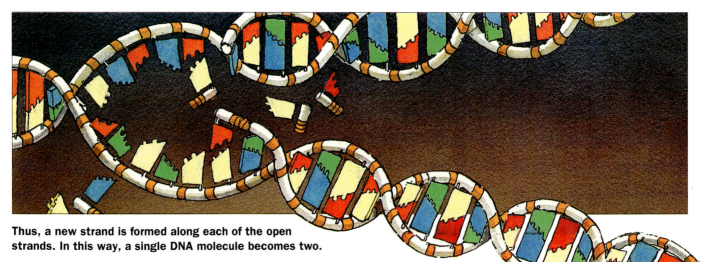

Thus, a new strand is formed along each of the open strands. In this way, a single DNA molecule becomes two.

How Enzymes Copy DNA

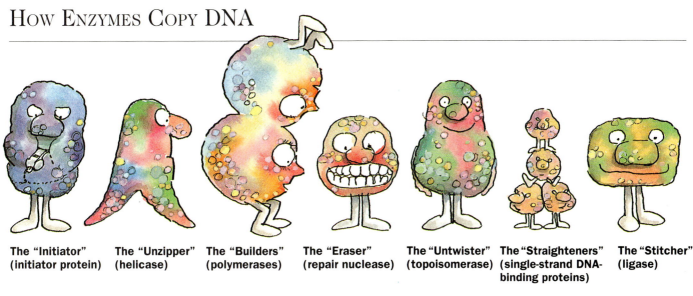

The "Initiator" (initiator protein)   The "Unzipper" (helicase)   The "Builders" (polymerases)   The "Eraser" (repair nuclease)   The "Untwister" (topoisomerase)   The "Straighteners" (single-strand DNA-binding proteins)   The "Stitcher" (ligase)

A Cast of Ingenious Characters

The sequence at the left oversimplifies. DNA doesn't copy itself any more than a recipe bakes a cake. DNA passively stores information. The team of proteins shown above does the actual copying, or replication. And they do it with an accuracy of only one mistake in every hundred thousand or so nucleotides!

DNA Replication — The Details

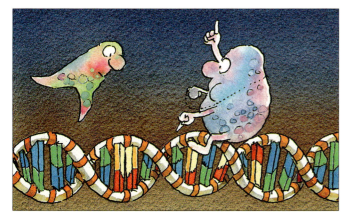

1. The initiator finds the place to begin copying and guides the unzipper to the correct position.

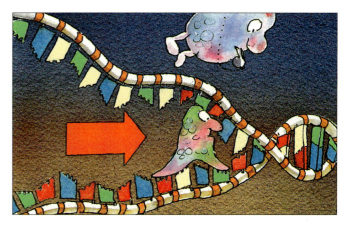

2. The unzipper separates the DNA strands by breaking the weak bonds between the nucleotides.

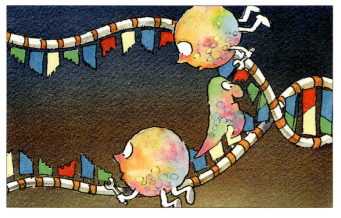

3. Then the builders arrive to assemble a new DNA strand along each of the exposed strands.

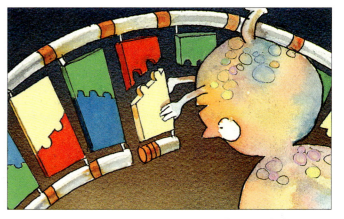

4. They build by joining individual nucleotides to their matching complements on the old strand.

DNA Replication — The Details

5. Free-floating nucleotides bring their own energy. Remember ATP (see page 46)? There's also GTP, CTP, and TTP.

6. As each new nucleotide is added to the growing chain, its phosphate bond energy goes into making the new bond.

7. The upper builder follows behind the unzipper, but the lower strand runs the opposite way.

8. Yet the lower builder must build in the same *chemical* direction. She solves this by making a loop...

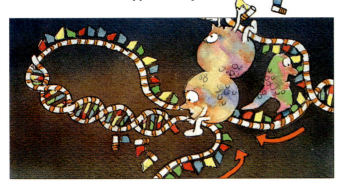

9. ...and building along the bottom half of it.

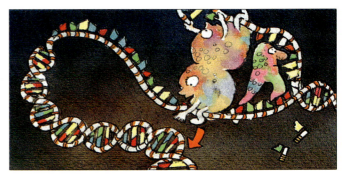

10. When she finishes a length, she lets go of the completed end...

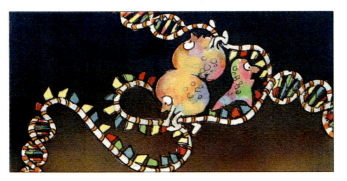

11. ...grabs a new loop, and continues linking nucleotides along a new stretch.

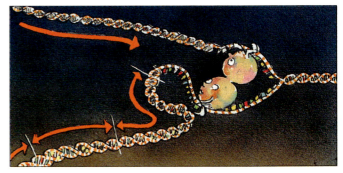

12. So, while the top new strand is built continuously, the bottom new strand is assembled in short lengths...

DNA REPLICATION — THE DETAILS

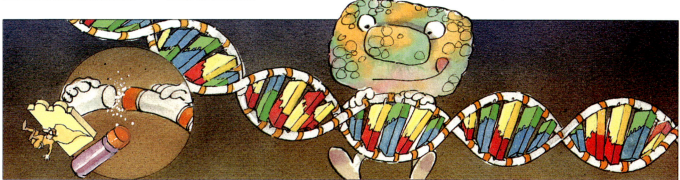

13. ...which are then spliced together by the stitcher. This reaction requires energy, supplied by ATP.

14. The straighteners keep the single DNA strands from getting tangled.

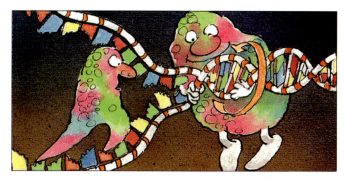

15. And the untwister unwinds the double helix in advance of the unzipper.

16. The initiator, the unzipper, the builders, the stitcher, the untwister, and the straighteners work together in tight coordination, making near-perfect copies at the rate of fifty nucleotides per second!

DNA Repair

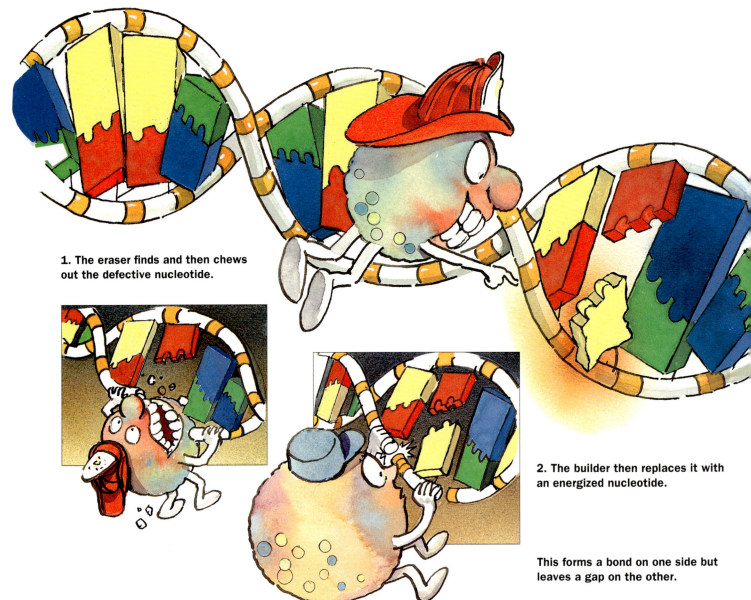

1. The eraser finds and then chews out the defective nucleotide.

2. The builder then replaces it with an energized nucleotide.

This forms a bond on one side but leaves a gap on the other.

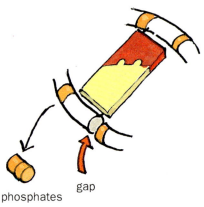

phosphates

gap

A Precise, Self-Correcting System

Although the system for copying DNA is extremely accurate, mistakes do happen; sometimes these mistakes can be devastating. Other threats to the integrity of DNA, which regularly damage nucleotides, include chemical events inside cells and ultraviolet light. The cell recruits an army of repair enzymes to handle these problems. Three kinds of repair enzymes regularly patrol DNA and repair any errors they find. First, erasers find poorly matching or damaged nucleotides and snip them out. Second, builders follow close behind to fill the gaps, using the other strand as a guide. Finally, stitchers restore the continuity of the backbone of the repaired strand.

Cells have evolved repair enzymes to help them survive those natural processes that regularly damage DNA. These enzymes continuously scan DNA and replace miscopied or damaged nucleotides.

Ahhh...

Here we go again.

Here's a close-up view of the ATP molecule donating its energy to make the bond.

3. The stitcher closes the gap using ATP for energy.

phosphate

ATP

95

DNA to RNA: Copying Genes into Messengers

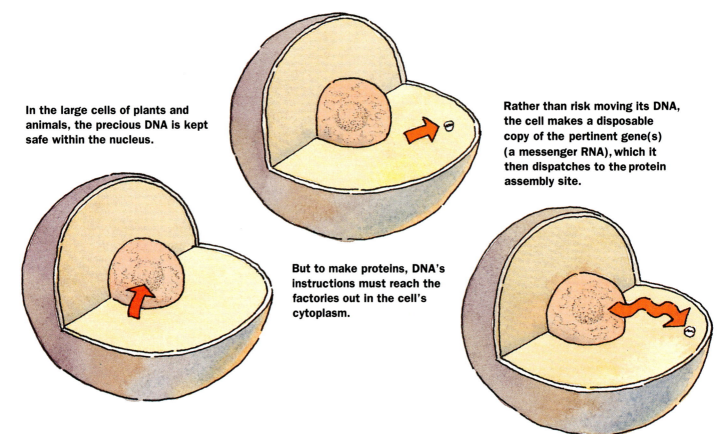

In the large cells of plants and animals, the precious DNA is kept safe within the nucleus.

Rather than risk moving its DNA, the cell makes a disposable copy of the pertinent gene(s) (a messenger RNA), which it then dispatches to the protein assembly site.

But to make proteins, DNA's instructions must reach the factories out in the cell's cytoplasm.

Transcription: Preparing the Daily Work Orders

While replication of DNA is the grand event preceding a cell's division into two, DNA also regularly participates in the daily business of living. As in our imaginary self-assembling computer, DNA's software provides the instructions to its hardware. These instructions get sent out from DNA's central storehouse in the cell nucleus to the protein-making assembly plants in the cell's cytoplasm (see Chapter 4) in the form of gene messengers, short stretches of information copied off the DNA. Messengers are made of a sort of throw-away version of DNA — good for limited work but not for long-term storage. Imagine going into a vault, taking out a set of precious instructions written on fine parchment, carefully copying the part you need on ordinary paper, returning the parchment to the vault, and then carrying the copy to the factory floor.

This process, called transcription, represents only the first stage of a larger operation dedicated to making proteins. You might notice (on the right-hand page) that transcription shares some of the mechanics of replication: DNA's double helix gets opened up, and a new nucleotide chain is built along a pre-existing strand that acts as a guide, or template. But the two processes differ. Transcription involves copying only one or a few genes at a time, not thousands. And the new, throw-away molecule that is produced is RNA (ribonucleic acid), a close cousin to DNA.

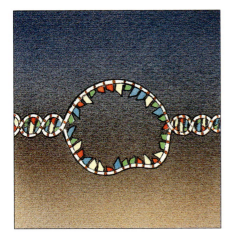

First, a small section of DNA is opened up.

One strand conveys the actual message of the gene.

The other strand acts as a template on which the messenger is made.

The messenger is made of nucleotides, similar to how DNA is built.

As the messenger is assembled, it separates from the template strand.

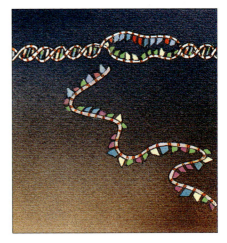

And when the entire gene is copied, the DNA releases the messenger.

Messenger-building requires the work of a single versatile enzyme.

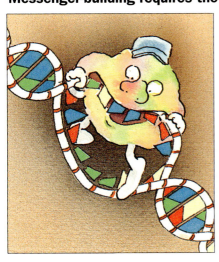

It finds the starting point along DNA...

...copies the gene...

...and then closes the double helix.

THE CHICKEN/EGG PROBLEM

The egg contains all the information needed to...

...make a new chicken...

...which grows up and...

A NEW WAY TO LOOK AT AN OLD PARADOX

Untangling the chicken/egg problem ("Which came first?") produces some real insight into the way life works. The paradox plays a trick by seeming to ask a single question when in fact it asks two very different questions at the same time. The first deals with cycles, the second with evolution. We need to separate these two.

We begin with the simple observation that any true loop has no beginning and no end. Chicken produces egg, egg produces chicken — in an endlessly repeating cycle. So the answer to "Which came first?" must be "neither." If you want to understand the underlying mechanism, however, try looking at the chicken as machinery and the egg as information. Machinery makes information, which instructs machinery, etc. But this, too, oversimplifies. While it's true that the egg has all the information needed to make the chicken, information by itself cannot do anything without some decoding machinery.

"A hen is only an egg's way of making another egg."
— *Samuel Butler*

...make all the proteins for an individual.

This DNA contains all the information needed, along with a little bit of chicken machinery, to...

98

...makes a new egg...

...and so on.

So we can say with more accuracy that the egg has all of the information plus just enough machinery to turn that information into living substance. In other words, every egg needs a little bit of chicken to go with it. The adult chicken, on the other hand, carries 100 percent of the information plus 100 percent of the machinery (that is, a complete chicken body); so producing a new egg is no problem.

The second question of the paradox might be rephrased as "Where did the chicken/egg cycle come from?" If we traced the ancestry of chicken and egg (both relatively recent "inventions") all the way back through billions of years, what would we find at the starting point? We can't be sure of the answer, but it may have been molecules that could function as both information and machinery (see page 182). From this beginning, chicken-ness would have arisen in tiny gradual steps.

Some of these proteins make more DNA...

...and so on.

DNA Packaging

BLOWING IN THE WIND

DNA has found a wealth of ingenious ways to package itself to ensure that its message will get to the next generation: pollen, nuts, seeds, spores, sperm, egg, etc. These vehicles often carry food with them to sustain the early phases of new lives. They also contain enough of the necessary machinery for DNA to get a new foothold — to express itself in the form of the next generation's protein molecules.

Most of these vehicles for DNA will get lost before they find the proper environment in which to develop. Their substance will be broken down into simple molecules, and their message lost. To ensure that this won't be the fate of all, life, profligate with energy and materials, makes millions of DNA carriers so that a few will succeed in getting their message through. However, sometimes even large numbers aren't enough to get the job done. Over eons of trial and error, the DNA of some kinds of organisms has found ways of using *other* kinds of organisms to help it pass its message down the generations. A plant's DNA, for example, instructs the plant's flower to produce nectar to attract bees or birds, which, in the process of nourishing themselves, not only ensure the survival of their own DNA but also pick up the flower's DNA-containing pollen and carry it to new locations. Think of the DNA of one kind of organism, then, as having the power to enlist the help of the DNA of another kind of organism to accomplish the crucial job of recreating itself in the next generation.

The human egg is this size. The DNA in its nucleus, if stretched into a single line, would be a yard long.

Chapter 4

MACHINERY

Building Smart Parts

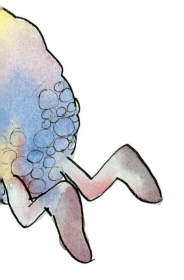

When we humans build a radio or a car or a computer, we assemble inanimate parts using the know-how we've accumulated over a few hundred years. When our cells build us, they use information accumulated over four billion years — and they build know-how right into the parts. The parts are "smart." Instructions in DNA are translated into many thousands of clever devices that do their tasks with astonishing fidelity, precision, and cooperation. Everything we do — think, laugh and cry, run and dance, conceive and give birth to children — emerges from the coordinated activities of a lively, intercommunicating society of protein molecules.

We call these proteins "machinery" because they move and thereby do work. Movement can be put to all kinds of clever tasks (see the next page). We call these proteins "smart" even though each one "knows" only a single trick (or, occasionally, two). By making a subtle shift in its internal structure, a protein can change its shape and change back again. If you watched one doing this all day, you'd likely be unimpressed with its I.Q. But if you watched several proteins, each performing its own task but working as a team, you'd begin to appreciate their cumulative "intelligence."

How does something as seemingly prosaic as DNA's long, monotonous sequence of nucleotides get converted into the 20,000 or so different kinds of protein molecules that perform the daily miracles in our bodies? That is the business of the cell's protein-making machinery.

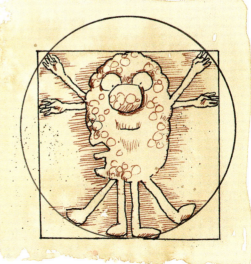

ABOUT PROTEINS

WHAT PROTEINS DO

Life's diversity can be traced to differences in the kinds and arrangements of protein molecules. More than half of the non-water weight of your cells is protein. Proteins do the daily business of living, giving cells their shapes and unique abilities. We've alluded to some of proteins' abilities earlier. Here's more about the key roles they play.

Enzymes

Enzymes are catalysts — they speed up the breaking apart and putting together of molecules. Their surfaces have special shapes that "recognize" specific molecules, similar to the way a lock accepts only a certain key. Enzymes themselves remain unchanged by the changes they bring about; so they can be used over and over again.

Transporters

Special transporter proteins in cell membranes function as tunnels and pumps, allowing materials to pass in and out of the cell.

Movers

Because the shape of protein chains is mostly determined by weak, easily broken and remade chemical bonds, these chains can shorten, lengthen, and change shape in response to the input or withdrawal of energy. The energy molecule ATP can activate one part of a protein molecule, causing another part of the same molecule to slide or take a "step." Subsequent removal of ATP causes the protein to return to its original shape, taking another step in the process. Then the cycle can be repeated.

Supporters

Long chains of folded or coiled proteins can form sheets and tubes — the cell's equivalent of posts, beams, plywood, cement, and nails.

Regulators

Enzymes that convert one chemical to another must do so in several steps. The first enzyme in a cycle "notices" when enough of the final product builds up and shuts down the assembly line. This ability to respond to feedback is built into the regulator's structure (see Chapter 5).

Defenders

Antibodies are proteins with special shapes that recognize and bind to foreign substances, such as bacteria or viruses, surrounding them so that scavenger cells can destroy them and flush them out of the body.

Communicators

To work together in harmony, cells must be able to pass messages back and forth. Proteins can act as cells' chemical messengers. Hormones are examples. Communicator proteins sit on the surface of the receiving cell to gather the incoming signal.

Grrrr...

Pumping Iron

Multiplyng Small Effects

Out of the 20,000 or more different kinds of proteins made in human cells, we have selected two — actin and myosin — to show how small molecular events can produce large effects. Actin and myosin are the proteins that make muscle work. Inside muscle cells, actin and myosin genes are translated into many millions of copies of each of these proteins. They line up to form a biochemical racheting device that uses ATP for energy to shorten and lengthen itself. This tiny molecular machine leads to the action of a bulging biceps through the simple means of scaling up. Millions of actin-myosin combinations are strung end-to-end in long fibers, and these fibers are bundled together into dense, parallel, elastic cables — the muscle cells. Each microscopic contraction of an actin-myosin combination is amplified into contraction of a cell. Collective cell contractions produce an overall grand contraction — the action of a muscle.

1. Actin molecules are long and thin; myosin molecules are thicker and have many "arms" and "hands" sticking out from their sides. The hands touch the actin molecules.

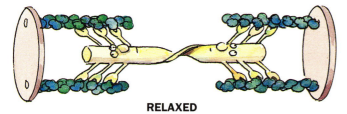

Myosin

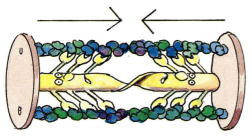

RELAXED

2. Each unit of contraction consists of two identical arrays of actins attached to discs and facing each other, connected by myosin. ATP binds to myosin's hands; in the process, phosphates break off, releasing energy. This causes the hands to grab the actin.

CONTRACTED

3. The subsequent release of the spent parts of ATP causes the arms to make a stroke like an oar, pulling the actins with their attached discs toward each other; this causes contraction.

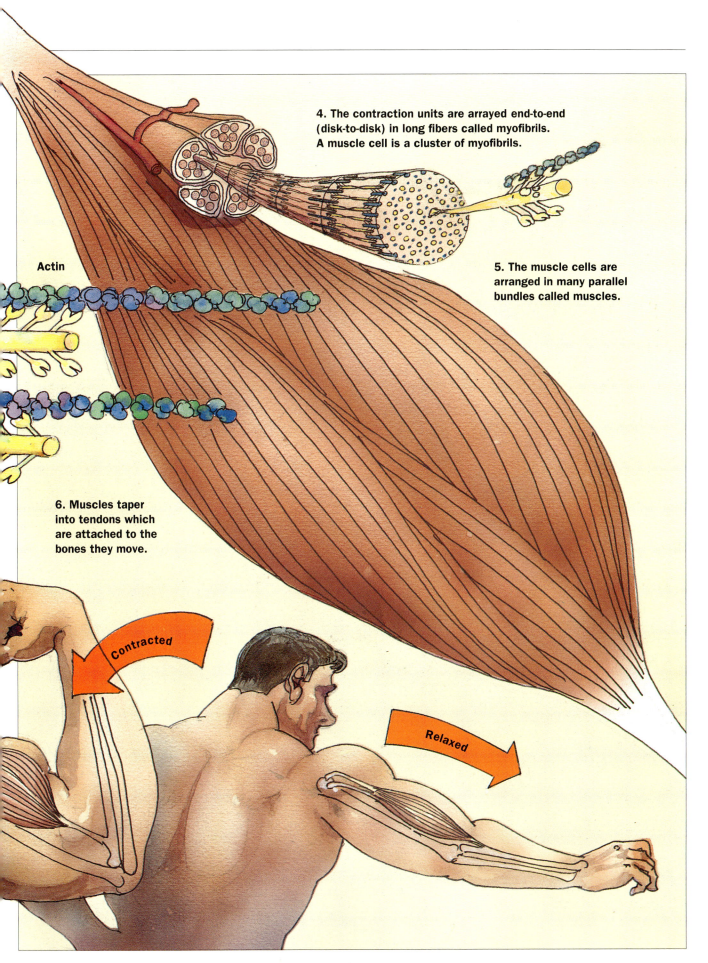

4. The contraction units are arrayed end-to-end (disk-to-disk) in long fibers called myofibrils. A muscle cell is a cluster of myofibrils.

5. The muscle cells are arranged in many parallel bundles called muscles.

Actin

6. Muscles taper into tendons which are attached to the bones they move.

Contracted

Relaxed

Sequence Makes the Difference

Underlying the bewildering variety of protein shapes and sizes is a surprising simplicity. When proteins are unfolded and stretched out, they turn out to be chains of amino acids. *The sole determinant of a protein's natural shape, and consequently its function, is the order of the amino acids in the chain.*

There are twenty — and only twenty — amino acids. Animals, plants, and bacteria use some or all of these amino acids in their protein chains. All amino acids contain carbon, hydrogen, oxygen, and nitrogen atoms, and two of them have sulfur atoms as well. Ten of the amino acids have electrically charged side groups that are attracted to water. These cluster on the surface of the folded-up protein chain where it's easier for them to make contact with the surrounding water in the cell. The other ten amino acids have no electrical charge and so tend to cluster on the inside of the folded-up molecule where they'll stay dry. The amino acids are linked to each other by strong covalent bonds between their backbone pieces (what we show as chain links). Once a protein is assembled, its amino acids form weak bonds with each other. These easily broken and reformed weak bonds give protein molecules their remarkable ability to change shape, which is the key to their functioning. They also give proteins great flexibility and mobility.

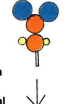

Each amino acid has a different side group with a unique chemical character...

...attached to a backbone piece that's the same for every amino acid.

When the backbone pieces are linked together in long chains, they become proteins.

It's the sequence of the amino acids that distinguishes one protein from another.

Protein Folding

Proteins find themselves mainly in one of two environments — water or fat. This explains why proteins fold the way they do. A protein in a watery environment folds its fat-liking amino acids tightly inside itself while its water-liking amino acids face the surrounding water. Proteins that reside in membranes, which are made of fat, do the opposite. Proteins can't do their work unless they're folded up correctly.

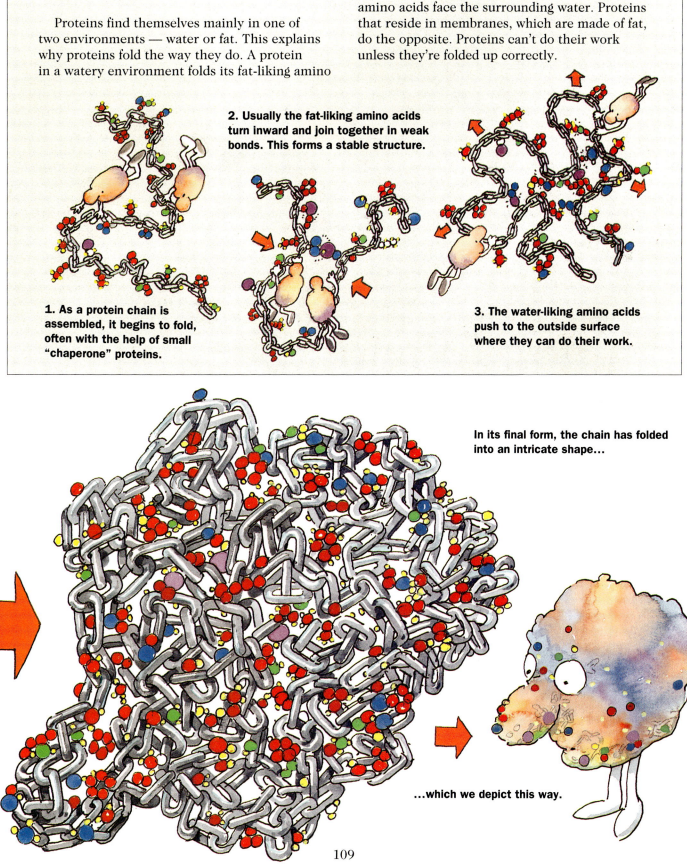

2. Usually the fat-liking amino acids turn inward and join together in weak bonds. This forms a stable structure.

1. As a protein chain is assembled, it begins to fold, often with the help of small "chaperone" proteins.

3. The water-liking amino acids push to the outside surface where they can do their work.

In its final form, the chain has folded into an intricate shape...

...which we depict this way.

Clothespins and Donuts

DoNutArama, a popular donut shop, makes twenty kinds of donuts. The donuts are so good that people buy big boxes of them. And each customer is very particular about having exactly the right kinds of donuts in exactly the right order in the box.

At first, the clerk at the counter tried shouting the orders to the kitchen staff, but they made too many mistakes. Written orders were out because the employees couldn't read the clerk's handwriting. Then someone remembered the colored clothespins in the basement. Maybe the clerk could somehow use the clothespins to transmit orders for donuts to the kitchen.

The clothespins came in four colors. The donuts came in twenty varieties.

What's the most efficient way to use four units to represent twenty units? The clerk worked out a *code*.

He first tried using combinations of two colors of clothespins: i.e., red + blue = jelly; yellow + red = chocolate; etc. He soon realized that there weren't enough different two-color combinations to represent all twenty donuts. But a three-clothespin code could produce sixty-four (4 x 4 x 4) possible combinations — more than needed for twenty different donuts. So he and his staff worked out and memorized a three-color code: red + blue + yellow = jelly; yellow + red + green = chocolate; etc. As the clerk took the orders, he put the correct color sequence on the line. In the kitchen, the decoder read the code, then hung the proper donut on the hook next to it. The packager took the donuts off the hooks and put them in their proper sequence in the box. Counter orders were transcribed into clothespin sequences and decoded into boxes of donuts, and things worked sweetly ever after.

Four different clothespins, taken three at a time, code for twenty donuts.

Jelly Plain Glazed Carrot Sugared

Coconut Maple Chocolate Carob Lemon

Sprinkles Nutty Blueberry Raspberry Pineapple

Custard Banana Marshmallow Almond Prune

Packager

How DNA Information Translates into a Working Protein

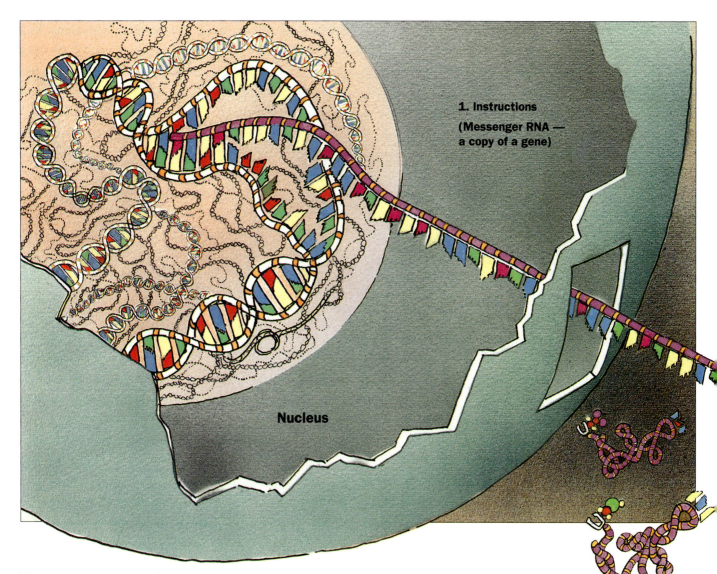

1. Instructions

(Messenger RNA — a copy of a gene)

Nucleus

Nucleotides and Amino Acids

A DNA molecule is many, many nucleotides (clothespins) long. It is composed of genes, which are, on the average, some 1,200 nucleotides long. Within each gene, the nucleotides are ordered in about 400 groups of three nucleotides apiece. Each nucleotide triplet gets translated into one of the twenty amino acids (donuts). The entire gene will be translated into a protein molecule that is about 400 amino acids long (the packaged donuts).

Here's how you make a protein. First, copy the sequence of nucleotides in a gene into a single strand of RNA (see Chapter 3, page 96) called messenger RNA (mRNA). Second, attach amino acids to small RNA molecules called transfer RNAs (tRNAs), or *adaptors*. These act like the decoder with her donut hook. Each adaptor recognizes a particular three-letter code. Third, bring the adaptors with their attached amino acids and the messenger RNA to a protein synthesis factory called a *ribosome* (the packager), which links up the amino acids to make the protein.

2. Adaptors

(Transfer RNA molecules with amino acids attached)

A transfer RNA is the key "decoding" unit between information and final protein product. Each has a three-letter code at one end and an amino acid at the other end.

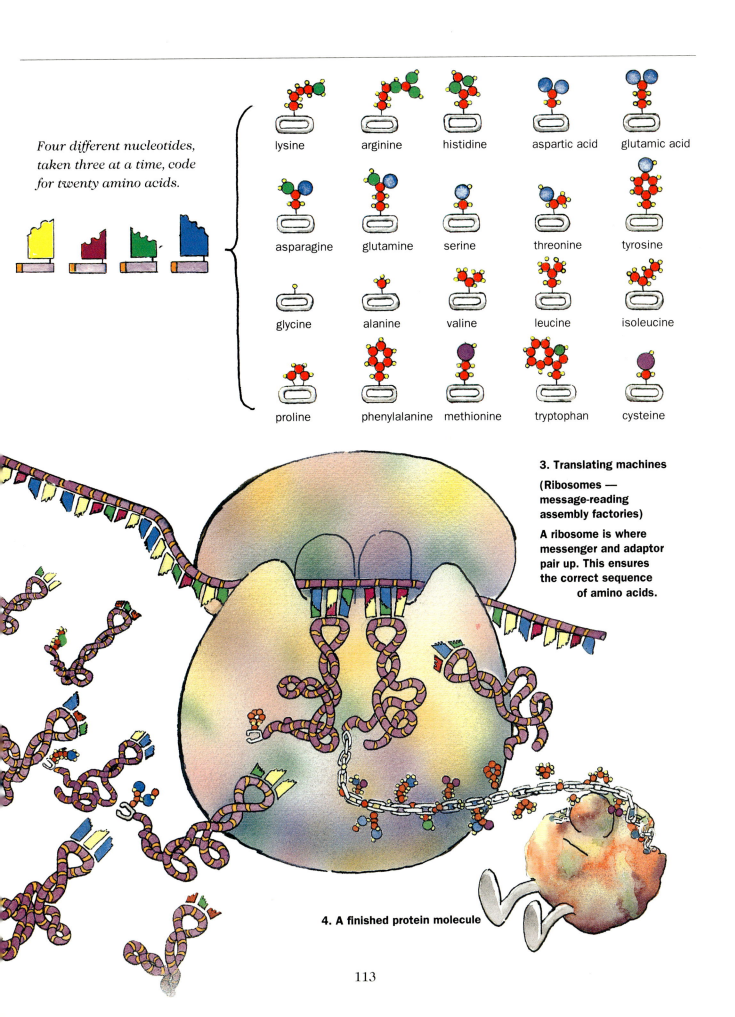

Four different nucleotides, taken three at a time, code for twenty amino acids.

lysine   arginine   histidine   aspartic acid   glutamic acid

asparagine   glutamine   serine   threonine   tyrosine

glycine   alanine   valine   leucine   isoleucine

proline   phenylalanine   methionine   tryptophan   cysteine

3. Translating machines

(Ribosomes — message-reading assembly factories)

A ribosome is where messenger and adaptor pair up. This ensures the correct sequence of amino acids.

4. A finished protein molecule

From DNA to Protein — A Multistep Process

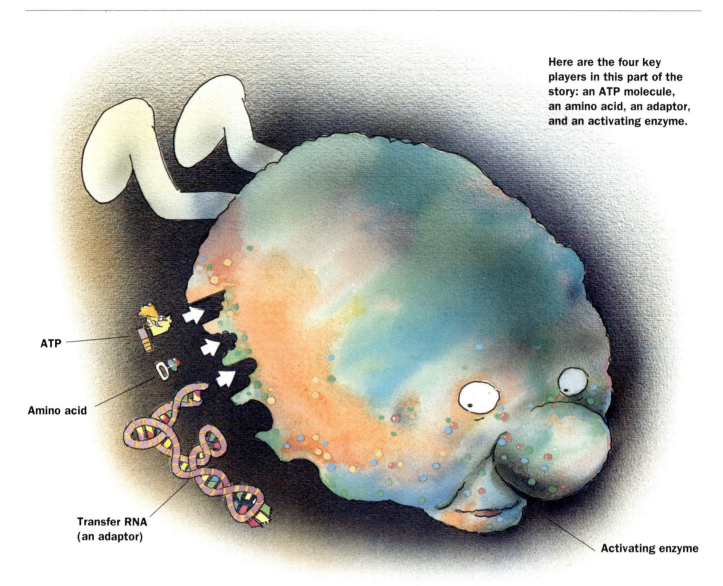

Here are the four key players in this part of the story: an ATP molecule, an amino acid, an adaptor, and an activating enzyme.

ATP

Amino acid

Transfer RNA
(an adaptor)

Activating enzyme

The Basic Idea

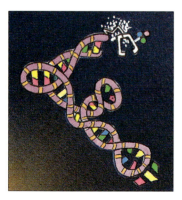

An energized amino acid gets put on an adaptor.

Charging the Adaptor

On the previous pages, we showed the translation process by which genes prescribe the order of amino acids in proteins. Now let's follow the key steps more closely. There has to be a chemical connection between each amino acid and each messenger RNA. Transfer RNA — the adaptor — makes that connection. One end of the adaptor carries a three-nucleotide code. This will match up with three complementary nucleotides on the messenger. A very smart enzyme, called an amino acid activating enzyme, energizes each amino acid and then attaches it — just the right one — to the opposite end of the adaptor. Since there are twenty amino acids, there must be at least twenty different activating enzymes and twenty different adaptors. In the panels on the right, we show the first steps in the construction of a protein: energizing amino acids and linking them to their adaptors.

114

The Details

ATP floats near the enzyme and docks in a place tailor-made for it.

Meanwhile, an amino acid floats into a dock nearby.

The two are brought closer together until...

...they bond...

...ejecting two phosphates from ATP.

The amino acid is now *energized*. (Note how the link is now open.)

Next, the odd-shaped RNA adaptor floats into view...

...and docks at another nearby site on the enzyme.

The end of the adaptor is brought closer to the amino acid until...

...the two are joined.

Energy flows into the new bond; the "spent" energy molecule is released.

Then the adaptor is released, with its amino acid attached.

Translation

Assembling the Protein Chain

An energized amino acid has been attached to one end of an adaptor, which carries at its other end a three-nucleotide code specific for that amino acid. Now the amino acid needs to be linked into a chain with others, in a specific order, to create a specific protein. This next phase requires the help of special machinery that can use the adaptors to "read" the nucleotide triplets on the messenger and assemble the appropriate amino acid chains. That's the job of the ribosomes. A ribosome is made of a larger and a smaller piece, each composed of about equal amounts of RNA and protein; it looks a bit like a designer telephone. The ribosome "reads" the tape-like message three units at a time, linking amino acids together as it proceeds. When it gets to the last triplet, which signals "stop," it releases the finished chain (the packager closes the donut box).

The Three Key Elements

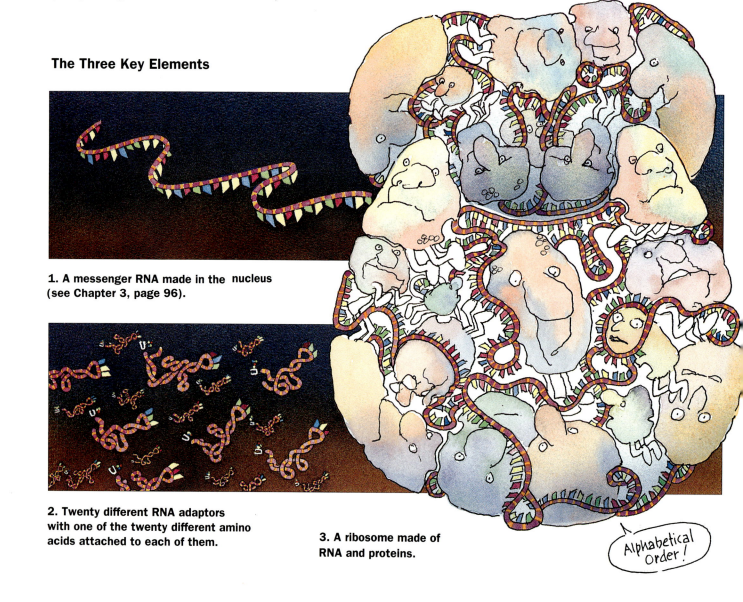

1. A messenger RNA made in the nucleus (see Chapter 3, page 96).

2. Twenty different RNA adaptors with one of the twenty different amino acids attached to each of them.

3. A ribosome made of RNA and proteins.

116

Assembling the Chain

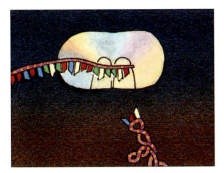

The messenger attaches itself to the smaller subunit of the ribosome.

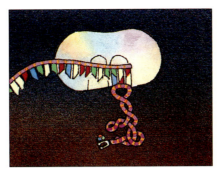

The first adaptor matches the messenger's first three nucleotides.

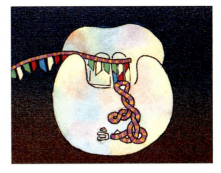

The larger subunit joins up with the smaller subunit.

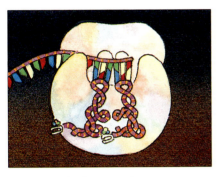

The second adaptor enters a second dock.

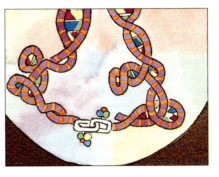

The backbone links of the first two amino acids join up.

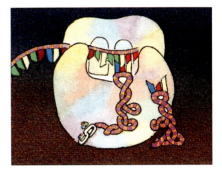

The messenger shifts to the right, and the first adaptor drops off.

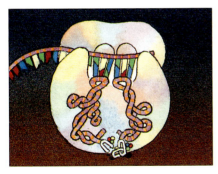

The next adaptor arrives at the second dock to add the next link.

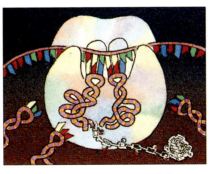

One by one, triplets are "read," and the protein chain grows.

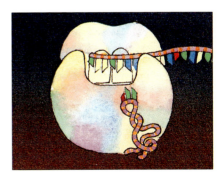

The final triplet signals "stop" — no adaptor fits it.

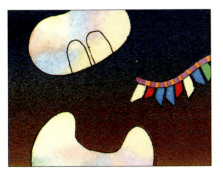

The ribosome separates and drops off the messenger.

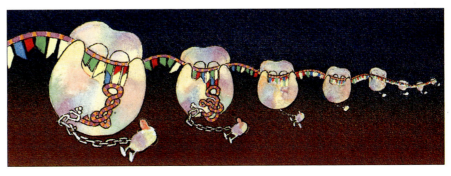

For efficiency, messenger RNA is read by more than one ribosome simultaneously.

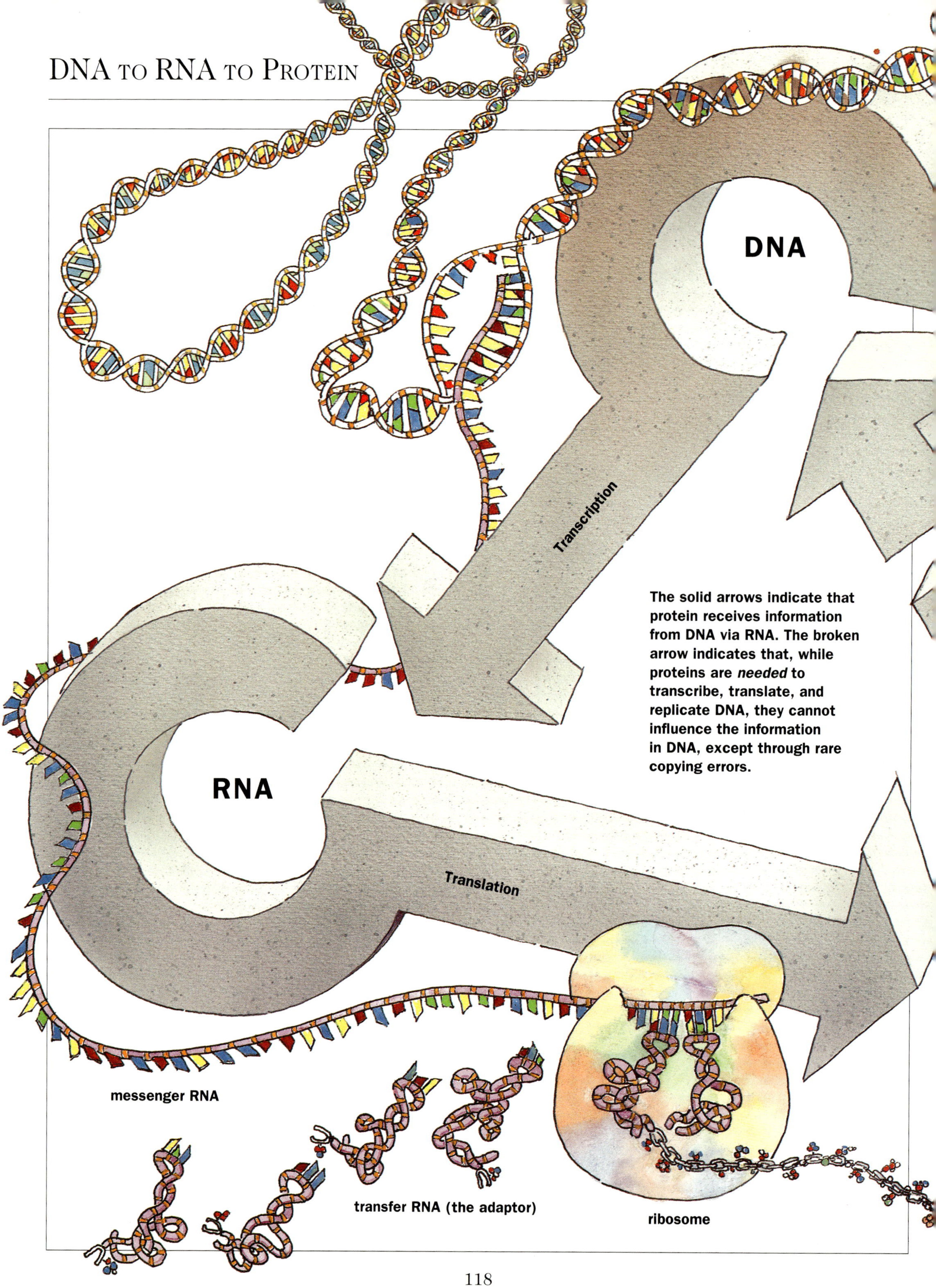

DNA

Transcription

The solid arrows indicate that protein receives information from DNA via RNA. The broken arrow indicates that, while proteins are *needed* to transcribe, translate, and replicate DNA, they cannot influence the information in DNA, except through rare copying errors.

RNA

Translation

messenger RNA

transfer RNA (the adaptor)

ribosome

118

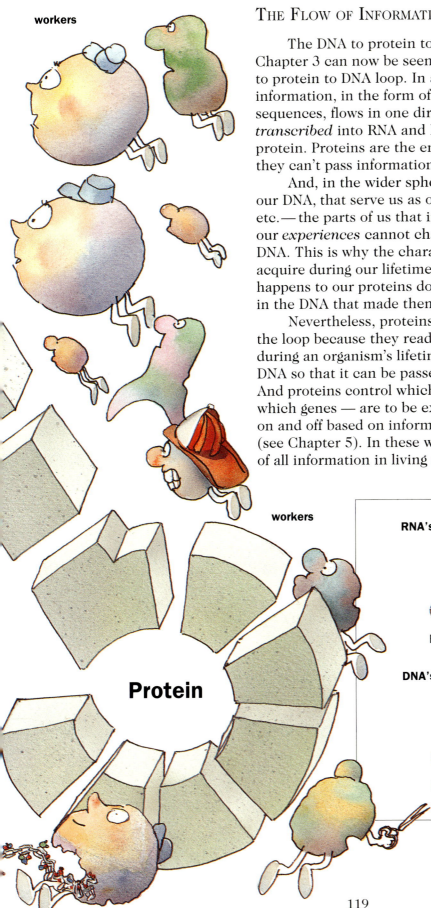

workers

THE FLOW OF INFORMATION

The DNA to protein to DNA loop we introduced in Chapter 3 can now be seen more accurately as a DNA to RNA to protein to DNA loop. In a strictly production-line sense, information, in the form of instructions written in nucleotide sequences, flows in one direction only: DNA's message is *transcribed* into RNA and RNA then gets *translated* into protein. Proteins are the end of the coded information line — they can't pass information back to DNA.

And, in the wider sphere, since it is our proteins, not our DNA, that serve us as our eyes, ears, noses, skin, nerves, etc.— the parts of us that interact with the world we live in — our *experiences* cannot change the coded sequences in our DNA. This is why the characteristics and behaviors we acquire during our lifetimes cannot be inherited. Whatever happens to our proteins doesn't change the coded information in the DNA that made them.

Nevertheless, proteins are keys to the continuity of the loop because they read and translate DNA's instructions during an organism's lifetime and are essential to copying DNA so that it can be passed to the next generation. And proteins control which parts of DNA's instructions — which genes — are to be expressed; i.e., they turn genes on and off based on information from their surroundings (see Chapter 5). In these ways, proteins influence the flow of all information in living systems.

Protein

workers

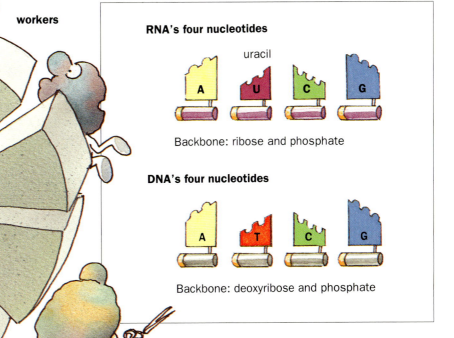

RNA's four nucleotides

uracil

A U C G

Backbone: ribose and phosphate

DNA's four nucleotides

A T C G

Backbone: deoxyribose and phosphate

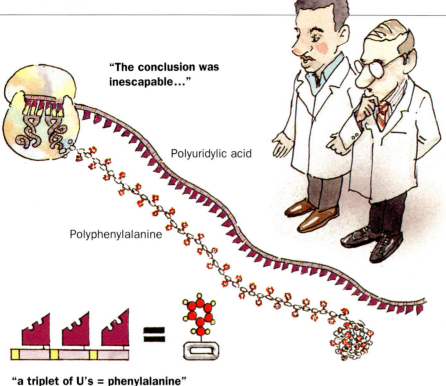

"The conclusion was inescapable…"

Polyuridylic acid

Polyphenylalanine

"a triplet of U's = phenylalanine"

How to Read the Genetic Code

The chart on the right summarizes the genetic code. Read it like a map with coordinates. Three nucleotides code for one amino acid. If you want to find the amino acid whose code is CAU, for example, find the box where C in the left-hand column meets A in the top row. This box contains histidine and glutamine. From this box look across to the right-hand column and find U. So histidine is represented by CAU. Note that it's also represented by CAC.

Life has used all but three of the sixty-four possible triplets that can be made using four nucleotides. So most of the amino acids are represented by more than one triplet. Three triplets don't code for an amino acid. Instead, these tell the protein-making machinery to stop.

CRACKING THE GENETIC CODE

In 1961, Marshall Nirenberg and Johann Matthaei, two young biochemists at the National Cancer Institute in Bethesda, Maryland, made an astonishing discovery. Not yet aware of the discovery of messenger RNA in Britain and France, they were searching for something like it: evidence that *some* type of RNA might program ribosomes to make protein. They took any samples of RNA they could lay their hands on and incubated them with ribosomes from bacteria, along with activating enzymes, ATP, transfer RNAs, and a mixture of amino acids. They looked to see if any of the RNAs stimulated protein synthesis. The results were not particularly encouraging until, by chance, they added an artificial RNA — polyuridylic acid (U-U-U-U…) — chains made of one nucleotide, uridylic acid, linked to each other as in natural RNA. Incredibly, the ribosomes obediently "read" the "poly U" chains into an artificial "protein," polyphenylalanine — long chains of the single amino acid phenylalanine! The conclusion was inescapable: The triplet code for phenylalanine must be UUU.

The exciting wider implication: If ribosomes can be induced to translate RNAs of any nucleotide sequence into protein, then RNAs of known nucleotide sequence could be incubated with ribosomes and watched to see what kind of amino acid sequence came out. Here lay the solution to the genetic code! Nirenberg and Matthaei pounced, as did others who learned of their discovery. A frenzy of experimentation ensued, with the result that all sixty-one of the triplet codes for the twenty amino acids were identified by 1965.

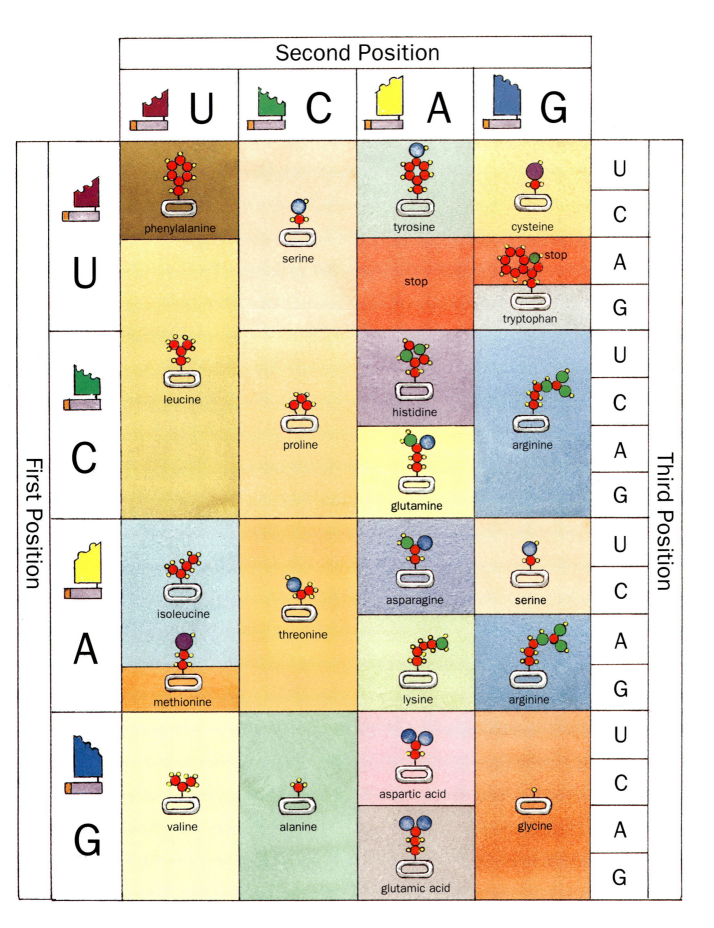

		Second Position			
First Position		U	C	A	G	Third Position

The Genetic Code table showing:

First Position U:
- UU: phenylalanine, leucine
- UC: serine
- UA: tyrosine, stop
- UG: cysteine, stop, tryptophan

First Position C:
- CU: leucine
- CC: proline
- CA: histidine, glutamine
- CG: arginine

First Position A:
- AU: isoleucine, methionine
- AC: threonine
- AA: asparagine, lysine
- AG: serine, arginine

First Position G:
- GU: valine
- GC: alanine
- GA: aspartic acid, glutamic acid
- GG: glycine

Third Position column: U, C, A, G (repeating for each row block)

THE UNITY OF BIOLOGY

A LIGHT THAT BLINKED...

Life first strikes us with its diversity. Evolution has filled every niche: Bacteria thrive in hot springs; fish plumb the depths of the sea; birds soar skyward, defying gravity. But when we look below the surface at the ways molecules work in cells, we cannot but marvel at their unity. All living creatures use DNA and RNA to store and replicate information, building them from the same four nucleotides. They make those nucleotides using very similar pathways. They translate nucleotide chains into proteins using the same twenty amino acids and the same genetic code. They use very similar translation apparatus — ribosomes, tRNAs, mRNAs, activating enzymes. If we take ribosomes from bacteria and put them in a test tube, they'll translate human messenger RNAs into human proteins — and *vice versa*. And many of the proteins — supporters, movers, communicators, transporters, and catalyzers — when dissected into their primary amino acid sequences — are quite similar in most creatures throughout the living world.

The realization dawns on us that we *all* had a common beginning. Billions of years ago, a tiny light blinked on somewhere and has come to illuminate every nook and cranny of our earth's surface.

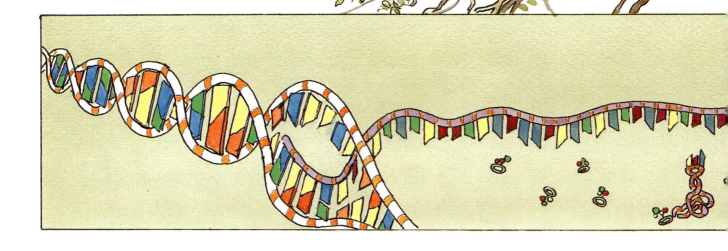

"We are educated to be amazed by the infinite variety of life forms in nature; we are, I believe, only at the beginning of being flabbergasted by its unity."

— *Lewis Thomas*

Our desired course is due north.
The plane has veered to the west.
We correct by turning toward the east.

Oops — we've overcorrected.
We need to steer to the west again.

Arriving at the desired destination
is a matter of many such corrections.

Chapter 5

FEEDBACK

Signaling, Sensing, and Reacting

Pilot, compass and steering mechanism all represent parts of a self-corrective circuit — a feedback control system (sometimes called a cybernetic system).

To an observer on the ground, an airplane appears to fly in a beeline toward its destination. But things look very different from inside the cockpit. Buffeted by winds or shifts in air pressure, the plane regularly drifts off course. When this happens, the pilot makes a correction by steering the plane in the opposite direction. If the pilot overcorrects, then he or she must correct the correction, and so on. The plane actually flies in a zigzag.

Feedback is a central feature of life: All organisms share this ability to sense how they're doing and to make changes in "mid-flight" when necessary. The process of feedback governs how we grow, respond to stress and challenge, and regulate factors such as body temperature, blood pressure, and cholesterol level. This apparent purposefulness, largely unconscious, operates at every level — from the interaction of proteins in cells to the interaction of organisms in complex ecologies.

How does feedback work? The process requires two elements: first, some kind of device to measure the difference between the current state of affairs and some preset "desired" state (like our pilot's compass and flight path); and second, some kind of responsive machinery that can reduce that difference (like the steering mechanism). The bigger the difference, the harder that machinery must work. This is negative feedback. But sometimes feedback operates as an amplifier, *increasing* the difference between the status quo and the objective. This is called *positive feedback* — and can lead to runaway and breakdown. It can also lead to creation and change, as we shall see.

ASSEMBLY LINES

THE AIRPLANE FACTORY

We can get an idea of how a living cell regulates itself by using an old-time airplane factory as a model. Skilled workers on multiple production lines assemble the essential components of each airplane from basic materials. Other workers stoke the fires under the steam boiler to generate the energy that powers the machinery. The bigwigs in the front office keep track of budgets, markets, and supplies, and they relay design and construction information. Floor supervisors, receiving instructions from the front office, control and coordinate how efficiently workers assemble the parts. From a distance, the production lines seem to hum along smoothly. Up close, things are a little more chaotic. Workers overshoot their production goals or miscalculate the number of parts needed; machinery breaks down. But the workers quickly correct for their mistakes, and the conveyor belts roll on.

The simplest living cell, though far more complex, is like the airplane factory, both in its organization and in its self-correcting behavior. Its workers are enzymes, teamed up in assembly lines. Some of these enzymes, the floor supervisors, possess the remarkable ability to evaluate the performance of the system and make the necessary adjustments. The cell's ultimate product is, of course, itself. It works to make more of its own components, to maintain them, to use them to further the whole organism's needs, and then to reproduce itself.

Both factories and cells are organized around a few basic rules:

1. Keep things moving in an even flow.
2. Don't allow components or products to pile up.
3. Be flexible and ready to respond to new demands.
4. Supervise every level of production.
5. Repair and replace machinery regularly.

CIRCULAR INFORMATION

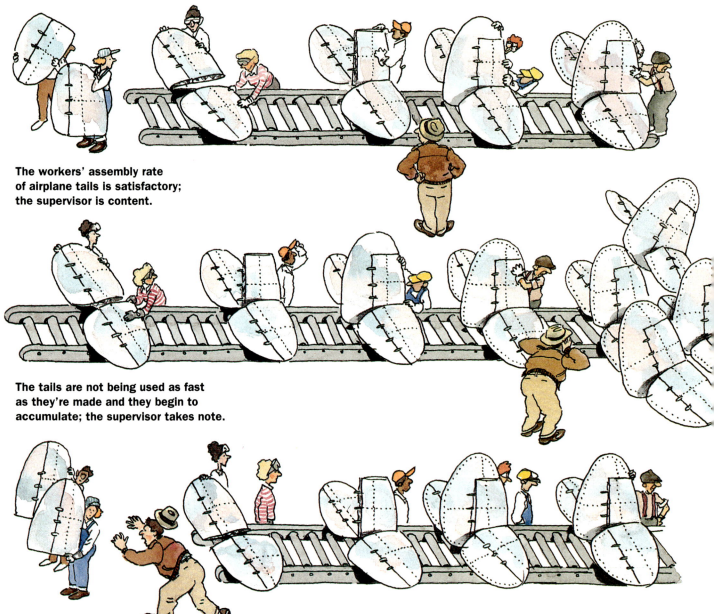

The workers' assembly rate of airplane tails is satisfactory; the supervisor is content.

The tails are not being used as fast as they're made and they begin to accumulate; the supervisor takes note.

The supervisor calls for a slowdown. The inventory of finished tails decreases as workers slow production. Later, as the supply of tails falls, the supervisor signals the workers to speed up their rate of assembly.

LINES AND LOOPS

An assembly line moves in one direction only — from input of raw materials to output of product, with the supervisor acting as the governor, or controlling agent. If too much product begins to accumulate, the supervisor slows down the input of raw materials. Conversely, if there are too many raw materials, the supervisor speeds up production.

To appreciate how feedback works, it helps to imagine the information (the signals that say "too much" or "not enough") as flowing in a loop. Bending the production line into a circle and stationing the supervisor at a strategic point overseeing both input and output gives him or her greater control. This arrangement is impractical for many factories, but it works beautifully inside cells, in molecular assembly lines like the one shown at the right.

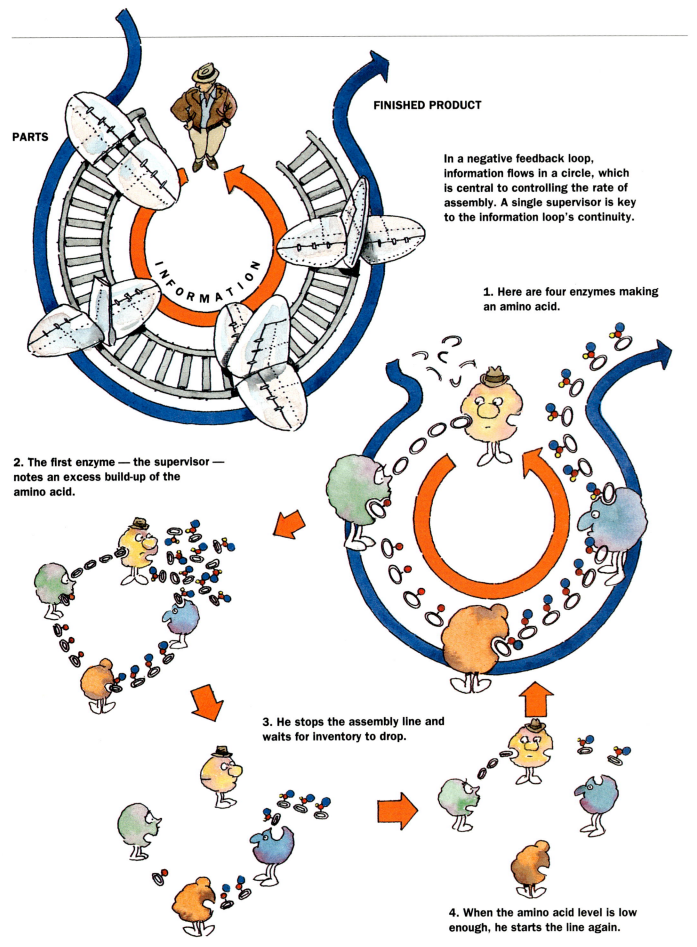

PARTS

FINISHED PRODUCT

INFORMATION

In a negative feedback loop, information flows in a circle, which is central to controlling the rate of assembly. A single supervisor is key to the information loop's continuity.

1. Here are four enzymes making an amino acid.

2. The first enzyme — the supervisor — notes an excess build-up of the amino acid.

3. He stops the assembly line and waits for inventory to drop.

4. When the amino acid level is low enough, he starts the line again.

ENZYME GOVERNORS

We can now better appreciate why we call enzymes "smart." Their unique chemical character gives them the ability not only to do their usual work of rearranging or disassembling other molecules (see page 104) but also to process information. Certain supervisory, or regulatory, enzymes do this by readily and reversibly *changing shape* in response to a signal. In addition to the working site on their surface where other molecules "dock" to get processed, regulatory enzymes have a *second* site specifically designed to hold a small signal molecule. Nestled in this special niche, the signal molecule acts like a finger on an on/off switch: It causes the enzyme to modify its shape so that its working site stops functioning. Allostery (literally "other shape"), the name given to this almost ridiculously simple behavior, underlies most of the unimaginably complex regulatory processes of life.

Regulatory enzymes are switched off — that is, they stop working — by literal contact with chemical information; they turn back on when that information (the chemical signal) is removed. (Some regulators operate in exactly the opposite way.) Like most protein behavior, these reactions are highly specific: For the most part, one and only one enzyme acts as a chemical switch for one and only one signal. But once a regulator's working site is turned off, it can shut down an entire production line. So a regulator can control a larger loop much as a governor controls the operation of a steam engine.

The behavior of allosteric enzymes offers a glimpse into the way living organisms attain their impressive complexity. As enzymes — the cell's workers — transport, build, and break down small molecules, allosteric enzymes — the regulators — control and coordinate these processes. And, as we shall see, higher-level regulators control subordinate regulators in a hierarchy of intercommunicating feedback loops.

Allostery — The Basic Idea

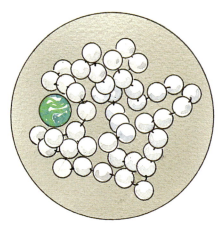

Imagine a tightly balled string of beads surrounding a marble.

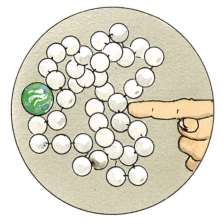

Now imagine pushing your finger into the beads on the side opposite the marble, causing the beads to shift.

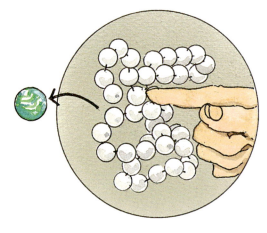

As your finger moves into its niche, the marble pops out.

Regulatory proteins behave in a similar way. When a signal molecule enters one site it changes the function of another site.

Shape one:
The active shape

In this shape, the working site is open — the system is "on."

Shape two:
The regulatory shape

In this shape, the working site is closed — the system is "off."

What exactly is the on/off signal in an enzyme assembly line? It is the final product molecule itself. This molecule is the equivalent of a message sent by the supervisor back to the first worker in the assembly line. When it fits into place on the enzyme, it says, "Enough already. Stop making us."

Site-filling by signal molecules is statistical. If lots of product molecules (signals) are around, a given set of identical sites are more likely to be filled.

As the concentration of product molecules decreases, they start falling out of the sites, leaving them empty.

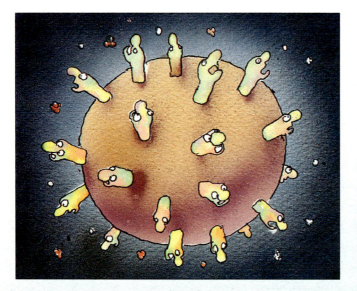

Allosteric receptors are embedded in the cell membrane.

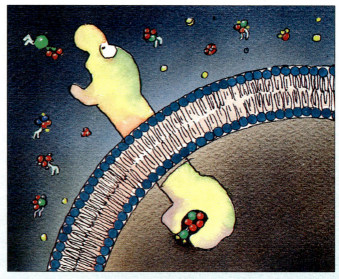

Each type of receptor matches a particular signal molecule.

PHEROMONES

Receptors inside nose

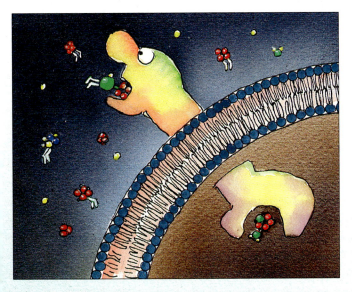

When a molecule fits...

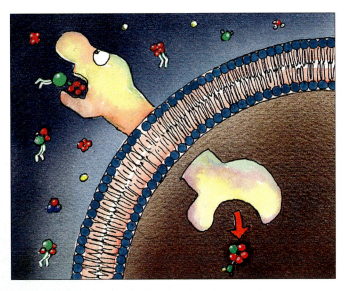

...the receptor changes shape, releasing an internal signal. This triggers cellular changes. (More on page 158.)

Ahh...spring.

PROTEINS ARE LIFE'S UNIVERSAL GO-BETWEENS

Allosteric mechanisms reveal a basic characteristic of life. Molecules tend to interact only when they have some chemical affinity for one another. But life's allosteric proteins can bring together molecules that have no direct chemical relationship. With such a protein acting as go-between, *any* small molecule can, in theory, act as a signal to influence *any* chemical process. A simple hormone molecule made in the brain — or thyroid, or ovary — can travel through the bloodstream to turn on chemical activities in cells all over the body. Pheromone molecules released into the air by a female mouse can reach receptor proteins in the nose of a male mouse and trigger a chain of events leading to mating. Allostery has enabled life to master the art not only of molecular control, but also of molecular communication. The vast evolutionary multiplication of chemical relationships brought about by regulatory proteins has created a network of interconnections within cells, between cells, and among the organs and tissues made up of those cells — a web of life.

A Local-Level Loop

A single assembly line is controlled by a regulatory enzyme, which is one of the members of the loop.

Local Control

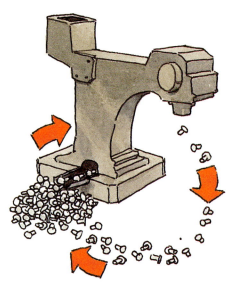

Here's an imaginary rivet-making machine. Its output is controlled by the accumulation of rivets.

When the machine makes too many rivets, they turn it off. This is local control.

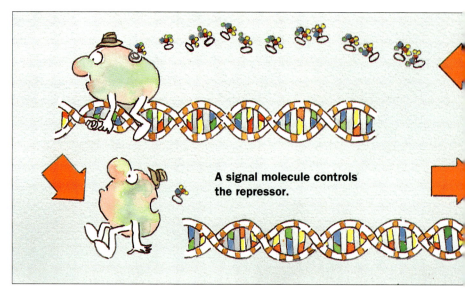

A signal molecule controls the repressor.

CONTROLLING THE MACHINERY THAT MAKES THE MACHINERY

We've seen how feedback control operates locally on the activity of each individual assembly line: The final product inhibits the activity of the first enzyme. This process is quick, sensitive, and reversible.

A higher-level process of feedback regulation controls the manufacture of the machinery that makes the product. This type of feedback control, which acts directly on the genes, is slower but far more consequential. It entails shutting down the whole process of making the enzymes involved in the assembly of a particular product — like laying off all workers on a particular assembly line.

A Higher-Level Loop

An allosteric protein known as a repressor, which has one site that responds to a signal molecule and another site that binds to DNA, controls the synthesis of the assembly line proteins themselves. When the repressor is active, it binds to DNA and prevents the synthesis of messenger RNA. (In some cases, the signal molecule makes the repressor active; in other cases, it makes it inactive.)

This higher-level loop includes repressor, genes, all the protein-making machinery, and the assembly lines themselves.

Higher-Level Control

Here's a rivet-machine maker. Like the machines it manufactures, it stops or starts depending on the number of rivets that accumulate. Too many rivets block the rivet-machine maker. This is higher-level control.

This system includes the big machine, the smaller rivet makers, and the workers and operators.

Here again, the product acts as the finger on the on/off switch; but in this case the protein that is switched on or off is a gene regulator protein that controls the production of one or more enzymes. By sitting on certain genes, the gene regulator (called a repressor) blocks the production of several enzymes (including the regulatory enzyme) that produce the product. It does so by stopping transcription of messenger RNAs (see page 96).

This second method of feedback control has the same purpose as the first: to avoid overproduction. But its effect are more far-reaching. It conserves materials and energy that the cell would otherwise use to make unneeded proteins, that is, those that make a currently overstocked product. This process is like controlling the conductor of the orchestra, instead of individual musicians.

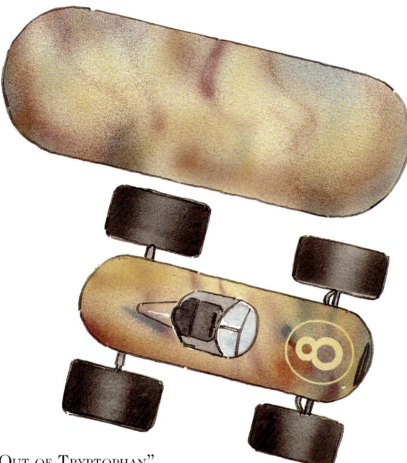

"OUT OF TRYPTOPHAN"

Imagine you're a bacterium, alone and naked in a vast liquid wilderness, trying to grow big enough to divide in half and become two of yourself. You're a lean, mean machine, like a stripped-down racing car — no extras for convenience, comfort, or luxury, but superbly adapted to a single purpose. And you achieve it with only about four thousand different kinds of proteins — as compared to the fifty thousand or more in a human cell. But your most notable achievement, the finest gift you've bequeathed to all higher-level organisms, is your capacity to orchestrate your own genes, using protein governors as switches. By expressing some genes while switching off others, you can adjust to an ever-changing world — a big evolutionary step for such a little guy.

Take your impressive ability to make all the amino acids and nucleotides you need using only sugar — which, after all, requires hundreds of separate enzymes to do (and which, by the way, no human can do). Normally, you scavenge amino acids and nucleotides from decaying organisms you find lying around. But in lean times, when the only thing available is sugar, you must improvise (i.e., make the enzymes that allow you to produce amino acids and nucleotides from sugar) or perish.

Here, for instance, is how you make the enzymes that make just one amino acid — tryptophan. The protein governor (here called a repressor) acts as a switch for turning a gene on or off.

Inside the Bacterium:

The genes are blocked when tryptophan is plentiful. ▶

There are five genes involved in making the five proteins needed for the tryptophan assembly line. When tryptophan is plentiful, an allosteric repressor protein binds to the DNA.

RNA polymerase, the enzyme that transcribes messenger RNA from DNA, can't bind to the DNA and transcribe it because the repressor is sitting on the binding site.

The genes become unblocked when tryptophan becomes scarce. ▶

When tryptophan molecules are scarce — e.g., when they're being vigorously consumed for protein synthesis — they no longer fill the repressor's regulatory site. The repressor loses its grip and falls off the DNA.

RNA polymerase is now able to transcribe messenger RNAs from the five genes. The messengers proceed to ribosomes where they are translated into five enzymes. These enzymes immediately get to work making tryptophan from sugar.

If tryptophan is not used rapidly enough, it will once more build up in the cell and reactivate the repressor, thus returning everything to the beginning of the control loop.

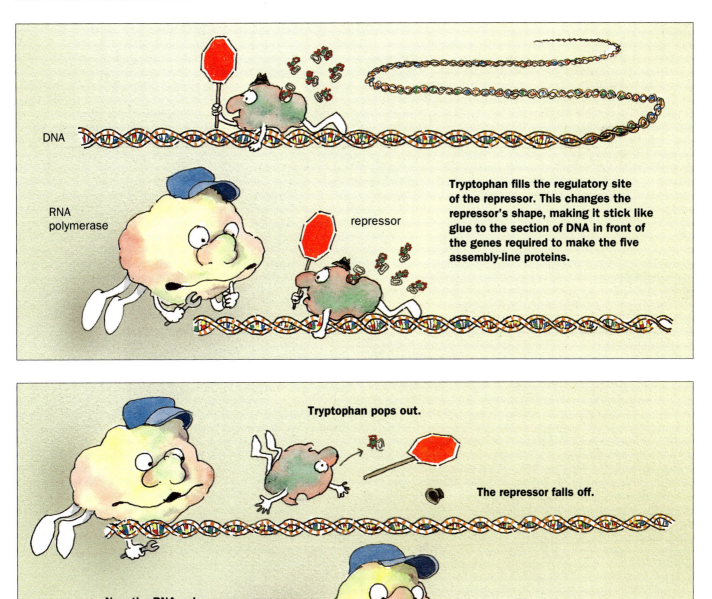

DNA

RNA polymerase

repressor

Tryptophan fills the regulatory site of the repressor. This changes the repressor's shape, making it stick like glue to the section of DNA in front of the genes required to make the five assembly-line proteins.

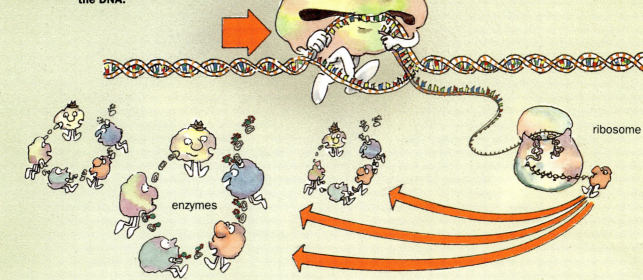

Tryptophan pops out.

The repressor falls off.

Now the RNA polymerase can copy the DNA.

ribosome

enzymes

RUNNING AND TUMBLING

The way bacteria find food in their environment — using *chemotaxis* (literally, "movement induced by chemicals") — is one of life's oldest forms of response to chemical signaling. A bacterium swims with the help of several flagella — long whip-like tails made of protein that are rotated by spinning disks, or "motors," on the bacterium's "skin." Counterclockwise rotation makes the flagella stream smoothly together to act like an outboard motor, propelling the bacterium straight ahead. Clockwise rotation makes the flagella flail about, causing the bacterium to tumble aimlessly. Normally, the direction of rotation reverses every few seconds so there's no consistent motion in any one direction. The bacterium runs a while, then tumbles a while. The result is a random "walk."

A bacterium's movement involves alternations of two kinds of action: "running" (top) and "tumbling" (bottom).

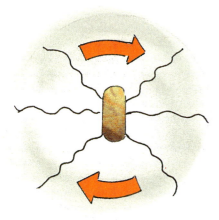

Each tumble changes the direction of the run. If it's not receiving any signals, the bacterium alternates runs and tumbles in (approximately) one-second intervals. This produces a random "walk."

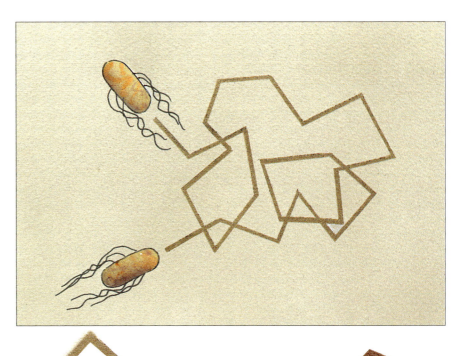

Food acts as a signal that makes the bacterium run more and tumble less as it moves toward higher concentrations of food molecules.

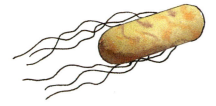

As long as its receptors indicate that it's heading toward more food, the bacterium does more running than tumbling. Its path, while still partly random, becomes more directed.

138

However, when the bacterium runs into food, it suddenly becomes more "purposeful." As food molecules fit into protein receptors on the outer surface of the bacterium, allosteric changes inside the bacterium signal the flagella motors to rotate counter-clockwise more frequently. The result is more running, less tumbling, always in the direction of *more* signal — more food. This preponderance of "directed" running over aimless tumbling continues as long as the number of food molecules hitting the bacterium's receptors keeps *increasing*. When the rate of increase levels off, the bacterium starts tumbling again, so it stays roughly in the spot where the concentration of food is the greatest. We can recognize in this tiny creature's remarkable ability to respond to chemical signals in its environment the rudimentary beginnings of purposeful behavior. The bacterium senses differences and then acts using its own internal energy to respond to that information.

food molecules

This self-correcting process, like a pilot's steering of an airplane, compensates for deviations of the bacterium's course away from the food.

FEEDBACK IN NEURAL CIRCUITS

1. Lock-step sequences:

At the simplest level, the spider follows a programmed sequence.

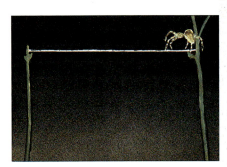

Release a strand into the breeze.

When it catches, cinch it up and attach the near end.

Walk across, paying out a second, looser strand. Attach at both ends.

Slide to the center of that strand and drop a third line, forming a Y.

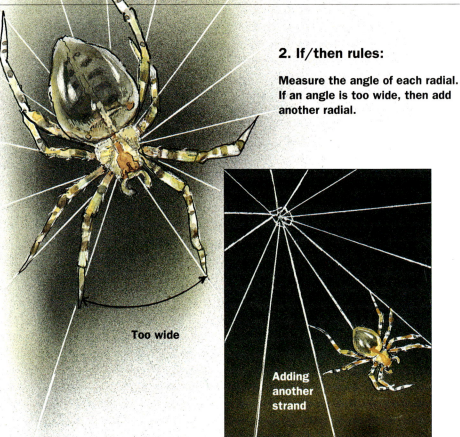

2. If/then rules:

Measure the angle of each radial. If an angle is too wide, then add another radial.

Too wide

Adding another strand

MAKING A FUTURE

Adjusting to the environment by going into the gene bank to make new proteins takes time. Complex organisms, particularly animals, need a quicker feedback system, both to respond to current conditions and to anticipate the future. The development of the nerve cell made this possible.

A spider spinning her web is attending to her future. If she chooses a good spot, if she makes her framework strong with uniform tension on the supports, if she spaces her orb lines evenly and makes them sufficiently sticky, chances are she'll eat well in the days ahead. The web-building spider is following a set of programmed rules embedded in her nervous system. More rigid rules dictate her specific behaviors; more flexible rules govern her overall strategies. Let's imagine you're a spider following four types of rules that we'll call (1) lock-step sequences, (2) if/then rules, (3) trial and error, and (4) simulation.

Here's a lock-step sequence: Cast a thread into the breeze and when it catches something, pull it tight and fasten the end nearest to you. Walk across this new span, paying out a loose second strand. Go back to the center of this second strand and drop a third strand straight down, forming a Y, etc. Lock-step sequences dictate an unvarying chain of specific procedures, like a soufflé recipe that must be followed to the letter for success. Here there's little room for feedback. The end point of one sequence is the starting point for the next.

At the next level, where feedback enters, are if/then rules: First, check each line for tension. If a line feels slack, then cinch it up. From the web's center, measure each angle between radial lines. If an angle is too great (meaning there's too much space between radials), then add another radial, etc. If/then rules, like lock-step rules, govern specific actions, but they permit modification based on feedback from sensory input.

A third type of rule doesn't specify an operation but says, more generally, "Repeat what works; stop what doesn't." Observers have noticed that if a spider's web is destroyed more than a couple of times while she's working on it, she'll abandon the job and start a new one. This is trial and error — or, better, trial and feedback.

The final type of rule is simulation: Rather than carrying out an elaborate process to see if it works, make a mental model of it and imagine the consequences. Such modeling requires more sophisticated nervous circuitry. If spiders can do it at all, it is in a very rudimentary way. They do in fact make a loose "sketch" of the final spiral pattern of a web; then, retracing their steps and using their legs to measure, they make a more precise version — and eat up the original "sketch."

All animals, and maybe even plants, work with a mixture of rigid and flexible rules. The more complex the animal, the more flexible the rules, which leads increasingly to the ability to modify behavior based on feedback from experience — a process more commonly called learning.

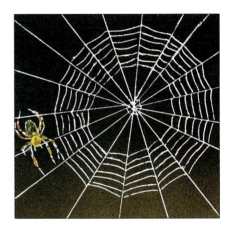

4. Simulation:

Make a preliminary "sketch" of the spiral using temporary webbing. Then, retracing your steps, make a permanent spiral, measuring carefully as you go. (Eat the original "sketch.")

3. Trial and error:

If the web sways too much in the wind, try adding weights to it. If this doesn't work, abandon the site.

CASCADING

POSITIVE FEEDBACK

Up to now, the only feedback we've discussed is negative feedback. *Positive* feedback occurs in a loop when a signal stimulates the production of more, not less, of a product. The more there is of the product, the more of it that is produced. Imagine an opera in which a certain note in the diva's aria acts as a cue to bring additional cast members onstage. As these new singers add their voices, still more performers come onstage. Thus, more brings on more.

When the diva sings a particular note...

...the lead tenor comes out.

Their two voices singing the same note...

...bring out a chorus of maidens.

Leads and chorus singing the note together...

...cue the lusty barbarians.

In biology the process of positive feedback is called cascading. Events trigger other events in ever-growing amplification. This can lead to a dangerous situation of "runaway," as in the case of addiction or the unrestrained cell growth of cancer. But cascading can also be a source of creativity, breaking a system out of the straitjacket of the usual and into something new. Like interest compounding itself, learning leads to more learning; success breeds more success. In the next chapter, we'll examine a vital cascade — the growth of an embryo from a single cell. And in the final chapter, we'll look at the grand cascade of evolution.

Each successive addition of voices summons an even larger contingent. Soon the stage is filled with singers.

ECOLOGY LOOPS

SELF-CORRECTING SYSTEMS

From a cybernetic perspective, an ecosystem is one huge feedback loop — a set of interconnected parts that act on one another so that a change in one part of the loop affects the other parts. In fresh water, for instance, fish eat algae and excrete organic waste; bacteria eat the waste and excrete inorganic materials; algae eat the inorganic molecules — each population thrives and multiplies in an interdependent cycle. Such balanced ecosystems have flexibility: Troublesome imbalances are corrected at some point within the loop.

A rise in water temperature may encourage an imbalance — an overgrowth of algae, for example. If the algae grow too dense, sunlight will fail to penetrate to their lower layers, and these layers will die. The resulting increase in organic waste will lead to explosive growth of the bacterial population, depleting the water of oxygen. Normally, the fish will restore the balance by increasing their numbers in response to the excess food — the algae.

We've seen how cyclical systems inside cells are controlled by allosteric proteins that act as supervisors, or governors. In an ecological system, the "governor" is usually the largest organism — the one with the slowest metabolic activity. The fresh-water ecosystem can't correct itself any faster than the fish are able to respond to the increased growth of algae. So even though ecosystems self-correct, they can be overwhelmed by sudden and extreme changes. Too much organic waste — in the form of sewage, for example — can totally deplete the oxygen in the fresh water and cause a collapse of the entire system.

This is a simple model. In most cases, ecosystems operate not as single loops but as networks of interconnecting loops in which both positive and negative feedback play a role. And if we could look deep inside those circuits, within each member, we would witness a never-ending making and breaking of molecules — the basic processes of life — presided over by myriad microscopic, bustling protein governors.

Water temperature rises, encouraging the overgrowth of algae. The bottom layer dies, increasing organic waste.

Bacteria multiply, depleting the water's oxygen content.

Fish multiply, reducing algae and restoring balance.

FISH

ALGAE

BACTERIA

INORGANIC NUTRIENTS

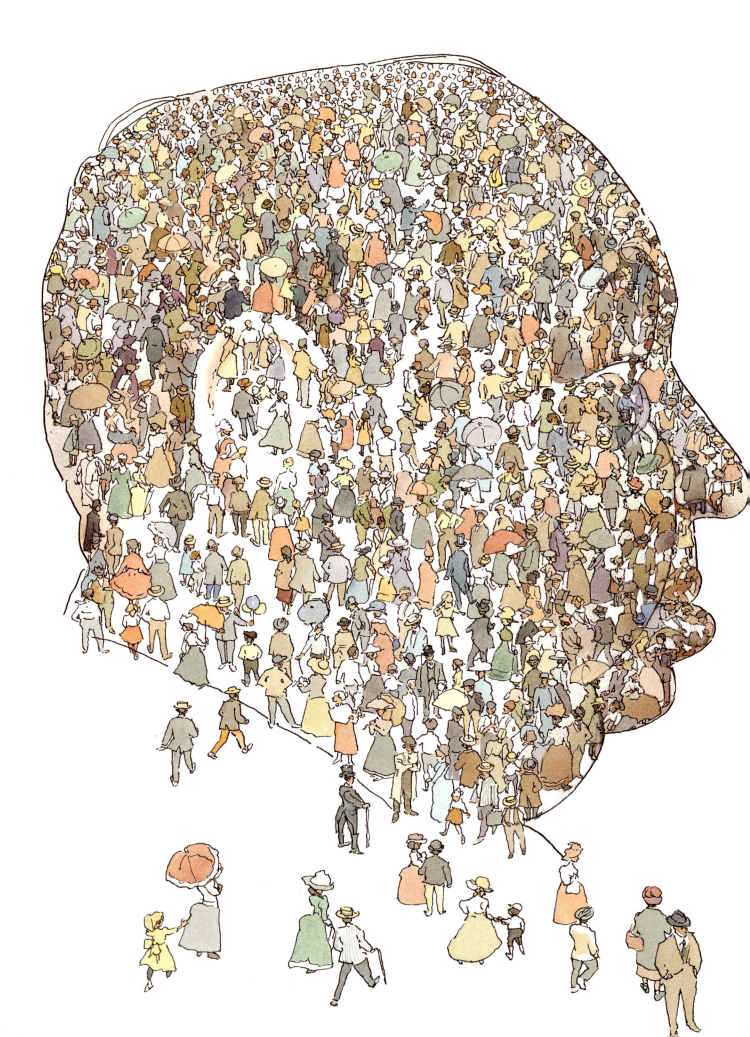

Chapter 6

COMMUNITY

E Pluribus Unum

Among the most astonishing revelations in biology was the discovery that all visible living creatures are themselves made up of living "creatures" called cells. Cells are not merely inert structural units or building blocks, but individual beings with lives of their own — living, reproducing, and dying just as we do. We are, in a sense, hives of cells. We move, eat, and speak thanks to the coordinated effort of specialized groups of individuals within our cellular community.

Cells are small. *Caenorhabditis elegans*, a worm so small it can barely be seen, is made of precisely 969 of them. You are made of an astronomical 5 trillion cells, give or take a few. Obviously, an aggregate that large requires an extraordinary degree of communication and cooperation. Cells must continuously "talk" to each other. They use electrical and chemical signals to control every action you take.

In community, cells organize themselves into specific patterns, maintaining their three-dimensional relationship with great precision. Two separated heart cells pulsating at different rates will, if placed side by side, fall into synchronous rhythm. Cells from different tissues will, if mixed in a blender, re-sort themselves into their original tissues in a short time.

Throughout evolution, the more cells tended to stick together, the more information they shared. Slowly, the rudimentary systems of primitive single cells became connected and elaborated on, ushering in higher-level behavior such as seeing, feeling, and thinking.

Consider a face: Its basic features change very little in the course of a year. Yet in that time, most of its original cells and all of the molecules of which those cells are made will be replaced by new ones. The fabric changes, but not the pattern.

Emergence

The Whole and the Parts

Life is more than the sum of its parts. If you throw a number of highly predictable individuals together, they will almost certainly interact and organize in completely unforeseeable and complex ways.

Take the information molecule, DNA. We've seen how DNA is a long chain molecule composed of four kinds of nucleotides (see page 85). There's nothing in the basic makeup or chemistry of those nucleotides that gives a hint of DNA's remarkable role in life. Only when they get strung together in the specific sequences of DNA do we perceive a wholly new quality — information. The true meaning of DNA lies not in its parts, but in the *organization* of those parts.

This truth holds for a whole host of phenomena, living and nonliving. For example, an individual water molecule by itself has no wetness. Wetness results when zillions of water molecules slide and tumble over each other, forming and breaking lattices. Individual atoms have no color. Color arises when atoms are organized in molecules, each absorbing certain light waves and transmitting others. A single brain cell (neuron) contains no "thought." Thought emerges as millions of neurons shuttle electrochemical impulses through ordered networks.

So the whole, rather than being simply the sum of its parts, is more like the *product* of its parts — a multiplication of the interactions among all the parts. In a true community, whether a plant, a human, or a city, individuals somehow manage to transcend themselves and become part of something much greater — even though each one still myopically works for its own ends and by its own simple local rules.

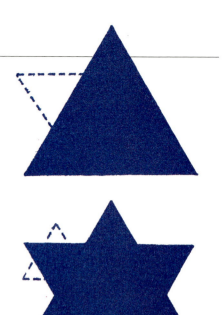

Emergent Patterns: When Simple Units Follow Simple Rules

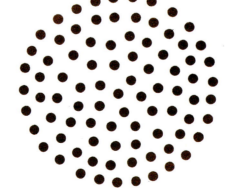

Imagine that each dot in this random assortment follows only two rules:

1. Keep exactly one dot-width from any neighbor.

2. Stay as close as possible to the center.

Emergent pattern: A disk

Rule: Add successively smaller triangles to the middle of each side of a triangle.

Emergent pattern: An increasingly intricate "snowflake"

148

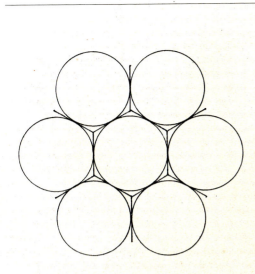

Rule: Assemble a group of spheres or cylinders so that they share walls, and use the least amount of material.

Emergent patterns: Hexagonal enclosures such as soap bubbles, some crystals, and bee honeycombs

Rule: Allow one of two connected initially parallel surfaces to grow faster than the other.

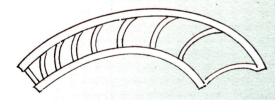

Emergent patterns: Rams' horns, plant tendrils, and chambered nautilus

THE EMERGENCE OF HIGHER-LEVEL BEHAVIOR

Community specialists —

How various leaf-cutter ants make a living

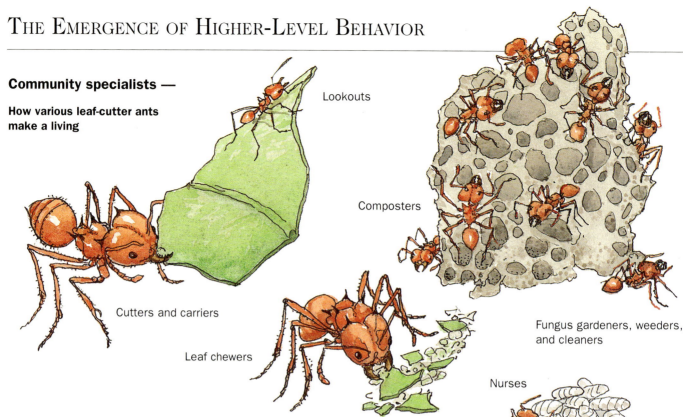

Lookouts

Composters

Cutters and carriers

Leaf chewers

Fungus gardeners, weeders, and cleaners

Nurses

Queen...

...and attendants

Community specialization has given social insects an enormous evolutionary advantage. While they represent only 2 percent of the world's insect species, they comprise more than half of the total mass of insect life on earth.

SUPERORGANISMS?

The social insects — bees, termites, and ants — provide one of the best examples of the way in which sophisticated behavior emerges from the interactions of simpler entities. Just as an organism "knows" more than any of its individual cells, an ant colony knows more than any single ant.

Ants, though nearly blind, display a remarkable sensitivity to chemical signals. They use these various substances to send simple messages, such as "follow my trail," "I'm a colony member," "on guard!" "help!" "I'm over here," etc. Biologist E. O. Wilson, a noted expert in ant behavior, suggests that an individual ant can probably send and receive some 15 different messages.

Imagine several scout ants out foraging for food. One stumbles upon some honey, and drawing upon her repertoire, she deposits a series of chemical messages saying "follow my trail" as she heads for home. The other scouts, having searched for food in vain, leave no returning trail. Sisters in the colony immediately pick up the trail left by the successful ant and go directly to the honey — each one reinforcing the trail on her return. Very quickly, a long column of ants makes its way directly to the food. They appear to follow each other, but in fact each one follows its nose (or, more accurately, its antennae), bumping into and stumbling over those returning with food. Notice that in this scenario, a random search quickly becomes an organized effort, even though each ant simply follows its own rules.

Clearly, the sharing of information brings the ant colony to a level of complexity (some even call it intelligence) not found in the individual ant. This is why some biologists refer to colonies of ants, bees, and termites as "superorganisms."

How to Build Without a Plan:

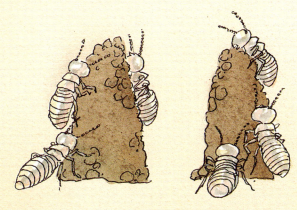

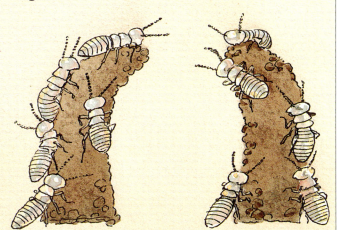

1. Individual termites randomly deposit dollops of mud and processed wood mixed with saliva that contains a chemical signal: "spit here."

2. Other termites add their deposits to the original ones.

3. Soon the little piles grow into columns. When the columns reach a certain height, the termites, still guided by a chemical signal, shift to a second message…

4. …"spit on the side nearest a neighboring column." Responding to this signal, the termites add new material to the columns so as to make them "bend" toward each other.

5. By following such local rules, termites can construct an elaborate layered network of arches and tunnels — their skyscraper nest — without the need for specific blueprints.

The Adventures of a Slime Mold: Individual to Aggregate to Individual

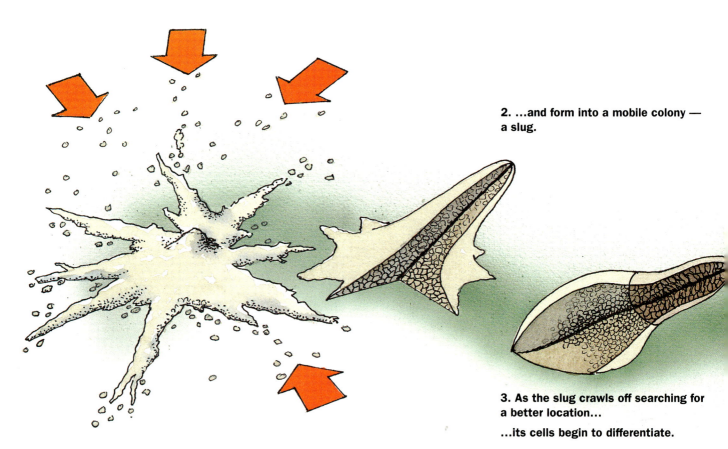

2. ...and form into a mobile colony — a slug.

3. As the slug crawls off searching for a better location...

...its cells begin to differentiate.

1. When food gets scarce, individual amoebae come together from all directions...

TWO-FACED AND SLIMY

According to the late author-philosopher Arthur Koestler, we all possess, in effect, two faces. With one face, we look inward and see ourselves as individuals. We express this face in our autonomy and independence. With our other face, we look outward and see that we are members of a larger community. We express this face in our communication and interaction. Having two faces is a matter not of choice, but of biology. Every living organism is both a whole unto itself and a part of something larger.

Few creatures exhibit this duality more dramatically than the strange, lowly slime mold. As free-living amoebae, residing on the forest floor, slime molds thrive on a diet of bacteria and yeast. But, when food becomes scarce, something unusual happens. A single amoeba, apparently self-appointed, begins to emit a chemical signal. Nearby neighbors, irresistibly drawn to the signal, "ooze" over and attach themselves to the signaler. Each new member of the cluster amplifies the signal by releasing its own signal (a good example of positive feedback, see page 142). More amoebae arrive, eventually forming a colony of up to 10,000 cells. Then a startling transformation occurs: The aggregate shapes

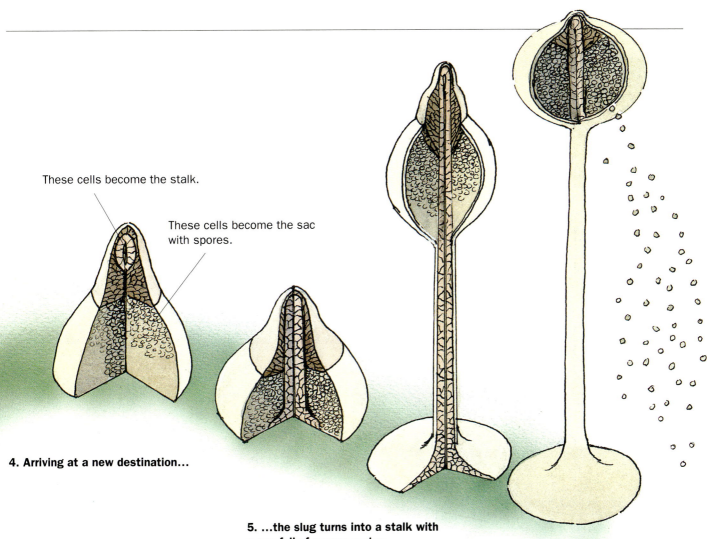

These cells become the stalk.

These cells become the sac with spores.

4. Arriving at a new destination...

5. ...the slug turns into a stalk with a sac full of spores on top.

6. Soon the dry spores are released, and each grows into a new, independent amoeba.

itself into a slug and begins to migrate to a new location, leaving a trail of slime behind it. As the slug moves, its cells differentiate into three distinct types, whose purposes become clear only when the slug arrives at a suitable new spot. One group of cells forms into a floor plate, or foot, with a tall, reedy stalk extending upward. A second group becomes a sac, which encases the third group, a cluster of spores. When these spores are eventually dispersed into the surroundings, they turn into new amoebae. And then the cycle begins anew.

The change from individual to communal member and back to individual echoes another primal cycle: egg to organism to egg again. In blurring the distinction between a part and the whole and in anticipating life's more complex reproductive strategies, the slime mold provides some tantalizing clues about how cells stick together, communicate, and differentiate — all key factors in the development of embryos in more complex organisms. The lowly slime mold demonstrates the awesome power of community, that is, how a group can cooperate to accomplish tasks unimaginable by single individuals.

A family builds a house in the middle of nowhere.

Later, using the same set of plans, the family's grown children build next door.

In time, the grandchildren add houses, using their own versions of the original blueprint.

By combining different parts of the building plans, some builders create new kinds of houses...

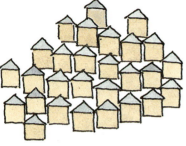

...and the cluster grows into a larger community.

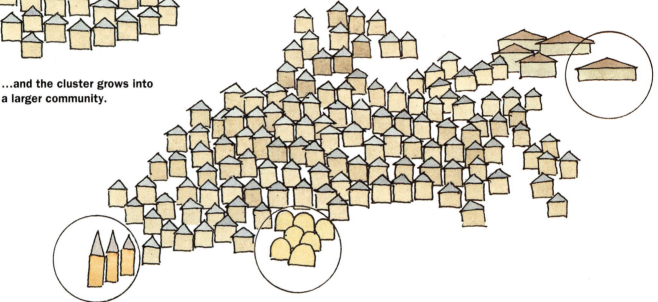

A SELF-ORGANIZING COMMUNITY

Few puzzles in biology have been more challenging than the question of how a single cell can grow within the space of a few days, weeks, or months, into a complex organism of millions, billions, or even trillions of cells.

We know that embryo development — life's building program — is based on information located primarily in the genes of each individual cell. We are still learning precisely how the proteins produced by these genes interact to coordinate such a complex event. It's difficult for us to envision a process in which so many things go on at once. First, cells grow and divide. Second, they differentiate, becoming specialists such as bone, skin, nerve, and all the other cell types in the body. Third, they migrate to various locations. And, fourth, they influence the behavior of their neighbors.

The simultaneous progression and interaction of these four activities lead quickly to extraordinary complexity. A growing cluster of houses offers a simple metaphor for this process.

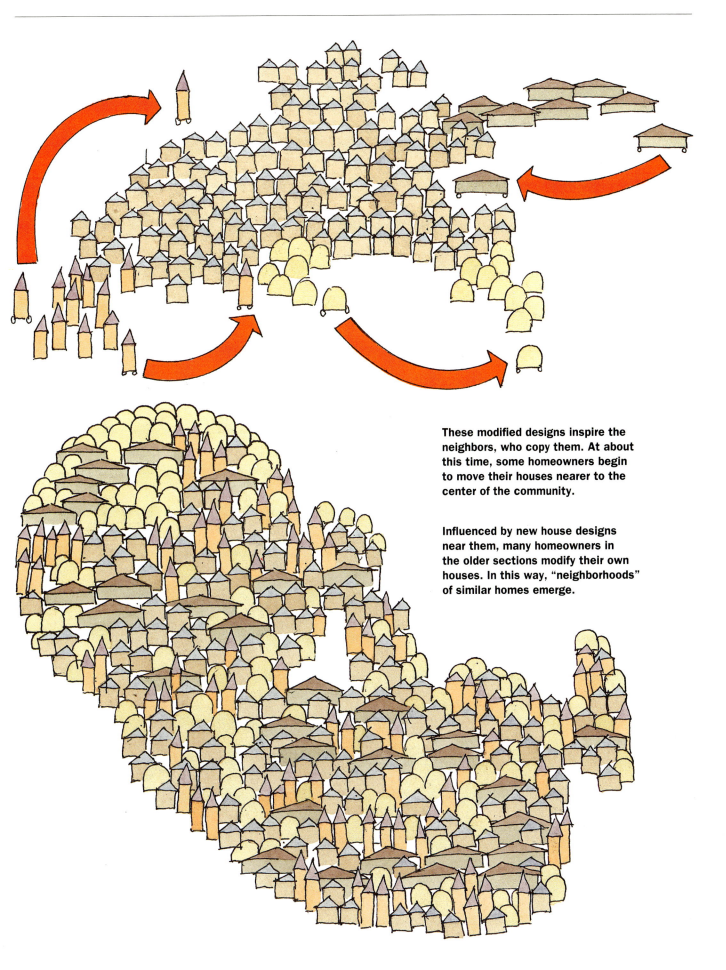

These modified designs inspire the neighbors, who copy them. At about this time, some homeowners begin to move their houses nearer to the center of the community.

Influenced by new house designs near them, many homeowners in the older sections modify their own houses. In this way, "neighborhoods" of similar homes emerge.

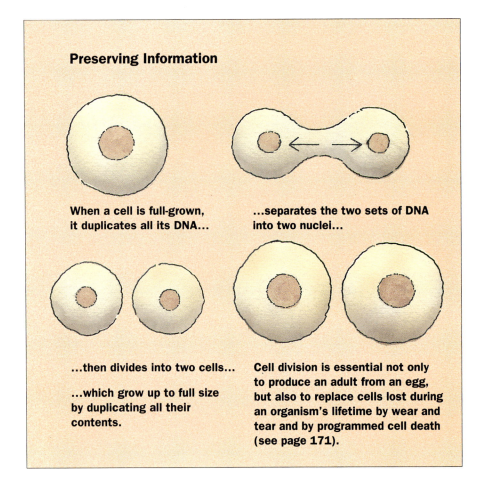

Preserving Information

When a cell is full-grown, it duplicates all its DNA...

...separates the two sets of DNA into two nuclei...

...then divides into two cells...

...which grow up to full size by duplicating all their contents.

Cell division is essential not only to produce an adult from an egg, but also to replace cells lost during an organism's lifetime by wear and tear and by programmed cell death (see page 171).

Forming a Hollow Ball

We each began as an egg fertilized by a sperm, a single cell. This cell divided, then divided again, and then again. Doubling over and over — 2 producing 4, producing 8, etc. — quickly leads to large numbers. If all these early cells divided at the same rate, it would take only about 30 divisions to make the many billions of cells of a newborn human.

Early on, cells divide but don't seem to change. By the time there are about a hundred cells, which takes about five days in human development, the cells have formed into a hollow ball. A cluster on one side will develop into the embryo, while the single outer layer will become the nourishing sac called the placenta.

DIVIDING TO MULTIPLY

Every cell of your body contains a complete set of all the information that went into building it. At first, this seems unnecessarily cumbersome. After all, wouldn't it be more sensible for a skin cell to contain only the information required for it to function as part of the skin? Why should it bother to carry information needed for brain cells or liver cells? If you were an architect planning to build an entire city, you wouldn't include in the blueprints for each building the entire set of blueprints for all the other buildings. But life does just that!

To understand *why*, it helps to know *how* cell division happens. All cells grow: They double their size by doubling everything they're made of; then they exactly double their DNA and divide in half. Two completely new cells replace each parent. Each of these new cells has received a complete genome — a complete set of genes — an exact copy of all the information in the parent. Enzymes perform this genome-doubling with great precision (see page 90). Evolution has clearly found this process preferable to some mind-bogglingly complicated mechanism for divvying up the genes during cell division. Each cell, endowed with a complete library, selects those books it needs for its particular purposes, and leaves the rest of the books on the shelves.

Gene doubling and apportionment into two cells, as discussed here, should be distinguished from what happens when eggs and sperm are made. This is a special case in which the mother's and father's sets of genes double as usual, then mix. Next, they divide into 4 cells, instead of two. These are sperm or eggs, carrying half of the parents' genes. (For more on gene mixing in sex cells, see page 201.)

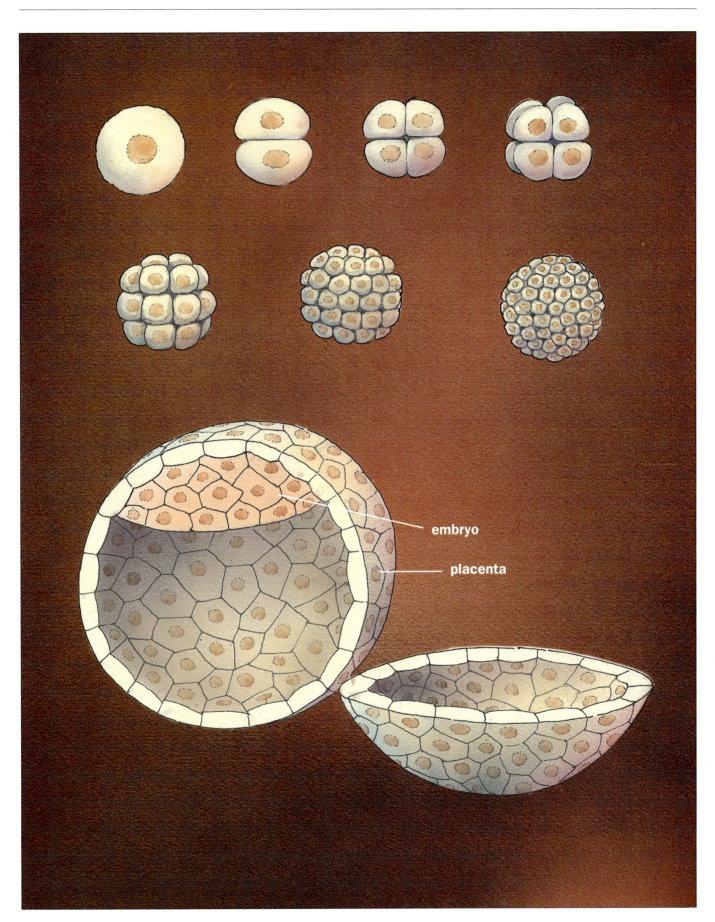

embryo

placenta

CELL SIGNALING

CASCADING INSIDE THE CELL

In Chapter 5 we asked you to imagine being a bacterium — a single cell, bounding independently through life, dividing to become two of yourself whenever conditions seem favorable. Now, try being one of the billions of cells of a multicellular organism. You still have inside you, in your DNA, the know-how to do anything a bacterium can do — and more — but now you have a role to play that is larger than yourself. You're one small part of a huge enterprise. And that means you're at the beck and call of signals. You can't grow and divide except in response to commands from somewhere else. These signals may be hormones, coming from a distant gland, or they may be proteins produced by neighboring cells. Your surface bristles with an array of receptor proteins, each responsive to a separate signal.

When a cell-division signal binds to a receptor's outer end, it causes the other end, inside the cell, to change shape (remember allostery, Chapter 5). This in turn switches on a rapid sequence of protein-to-protein contacts, which eventually activates the genes that turn on the machinery of cell division. (The energy for these contacts and activations is supplied by ATP.)

A major step in cell division, shown here, is the start of DNA duplication.

Such signal-induced sequences of molecular events resemble a cartoon in which a crowing rooster startles the cat who jumps off his perch causing a steel ball to roll down a chute, striking a switch that turns on the stove, heating up the coffee.

Many such protein-to-protein relays pass signal information to genes each time cells undergo any of the multitudinous changes needed to transform themselves from tiny blob into newborn baby.

Sometimes a cell turns "bad"; it becomes a sociopath, dividing at will, jostling its neighbors, and taking off for distant parts. This cell-run-amok is a cancer cell. Inside it, one or more of the signaling proteins have been damaged. Such damage, caused by a mutation in the gene responsible for making that particular protein, can have devastating consequences for the harmonious operation of the whole cell community.

All these relays are currently the subject of intense biological investigation. Unlocking their secrets will help us solve the mysteries of embryonic development — how life is re-created at each generation — and of cancer — how life can be destroyed by damaged relays.

An incoming signal...

...binds to a receptor...

...activating a messenger protein...

...which proceeds to the nucleus...

...and binds to a regulatory protein...

...which falls off a gene...

...starting up production of a messenger RNA...

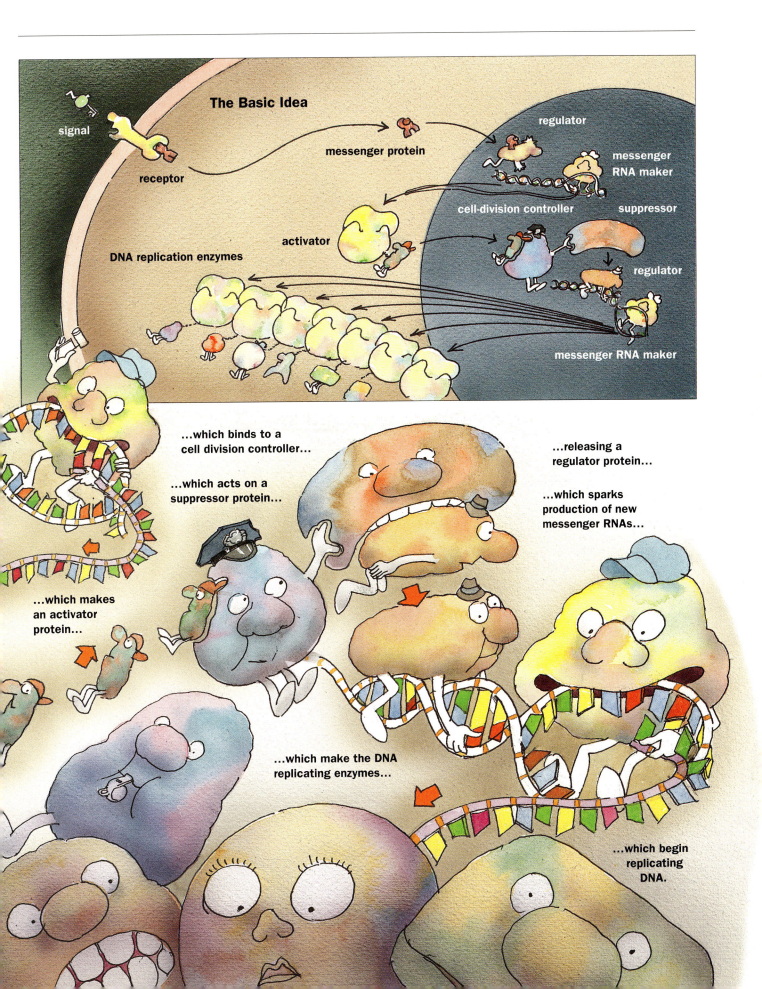

The Basic Idea

signal

receptor

messenger protein

regulator

messenger RNA maker

cell-division controller

suppressor

activator

regulator

DNA replication enzymes

messenger RNA maker

...which binds to a cell division controller...

...which acts on a suppressor protein...

...releasing a regulator protein...

...which sparks production of new messenger RNAs...

...which makes an activator protein...

...which make the DNA replicating enzymes...

...which begin replicating DNA.

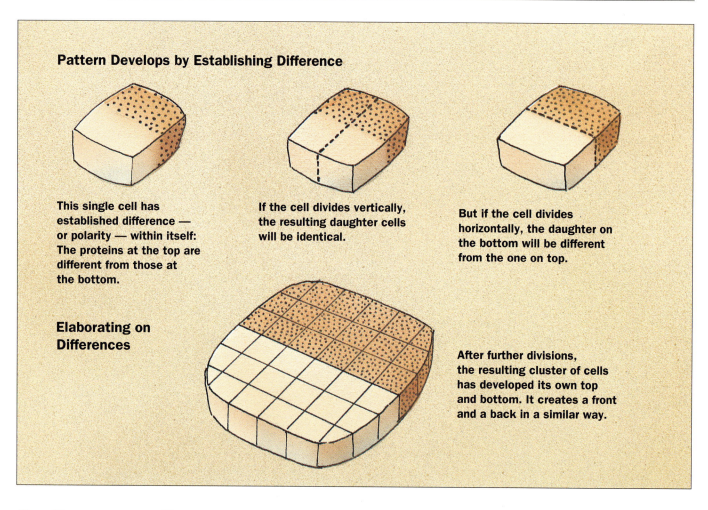

Pattern Develops by Establishing Difference

This single cell has established difference — or polarity — within itself: The proteins at the top are different from those at the bottom.

If the cell divides vertically, the resulting daughter cells will be identical.

But if the cell divides horizontally, the daughter on the bottom will be different from the one on top.

Elaborating on Differences

After further divisions, the resulting cluster of cells has developed its own top and bottom. It creates a front and a back in a similar way.

THE BEGINNINGS OF PATTERN

Scientists in the past, peering through their crude microscopes, swore they could see a tiny, fully-formed human figure huddled inside each human sperm cell. But, embryo development begins quite differently.

The first step toward the creation of a body comes when a few cells occupying a space no bigger than the point of a pin begin to take on the general character of what we'll call "topness," and others adopt "bottomness"; still others assume "frontness" or "backness"; "outsideness" or "insideness." There's no sign yet of a head or a tail, a backbone or belly, skin or internal organs — to say nothing of things in-between.

Multiplying cells that have made the commitment to, say, topness will, generation by generation, make small changes in their character and forge new relationships with their neighbors so as eventually to become a recognizable head. Each step depends on the specific changes that have occurred earlier. This is why we say "memory" is essential to embryonic development. Cells can't arrive at the final version of *what* they will be until their forebears have determined *where* they will be.

The Embryo Begins to Take Shape

On page 157 we showed the embryo as a cluster of cells inside a hollow ball. At the top right we show how this cluster begins to flatten out as a disk (top row) which then elongates and forms three layers: one for skin and nerve cells; one for the digestive system; and one for all other cells types. (The layers are shown separated in the diagram at lower right.) We use a simplified grid to portray the growing number of the embryo's cells.

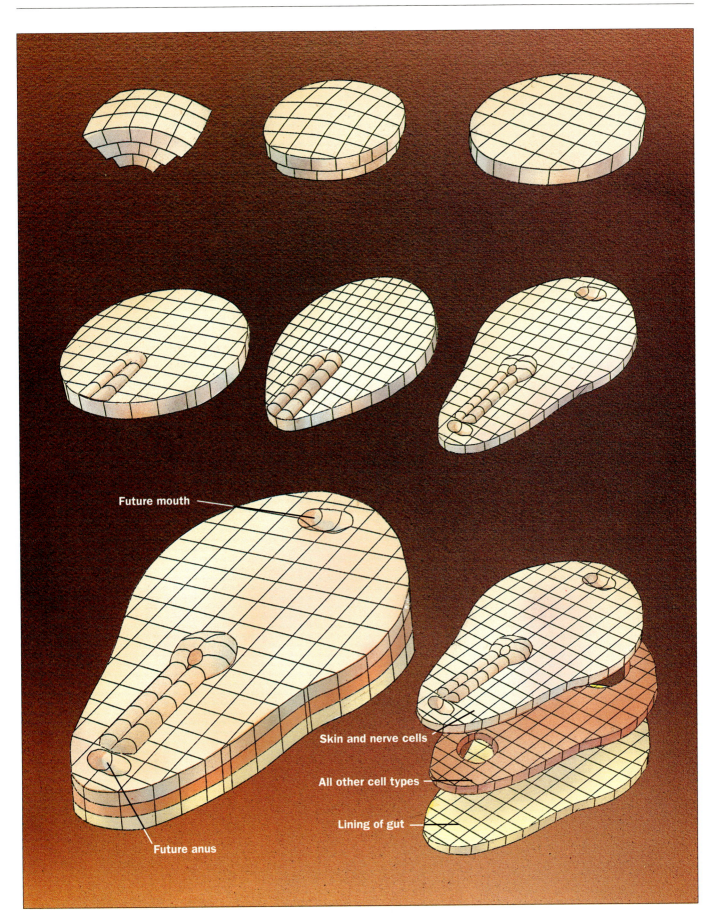

Future mouth

Future anus

Skin and nerve cells

All other cell types

Lining of gut

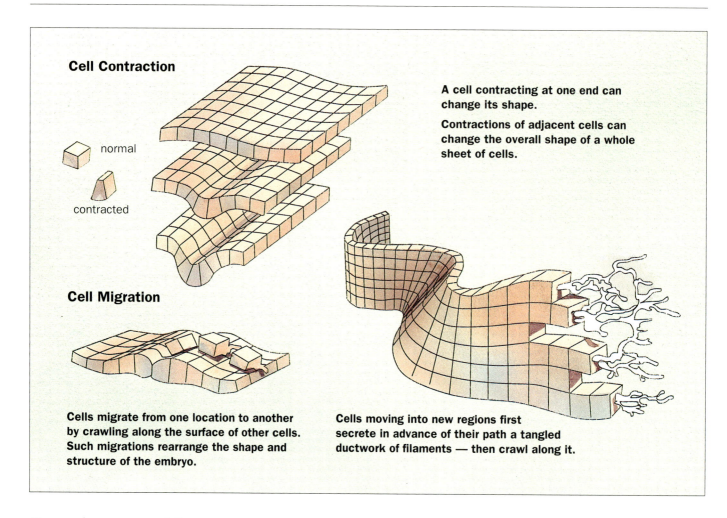

Cell Contraction

normal

contracted

A cell contracting at one end can change its shape.

Contractions of adjacent cells can change the overall shape of a whole sheet of cells.

Cell Migration

Cells migrate from one location to another by crawling along the surface of other cells. Such migrations rearrange the shape and structure of the embryo.

Cells moving into new regions first secrete in advance of their path a tangled ductwork of filaments — then crawl along it.

CONTRACTING AND MOVING

As the embryo begins to take shape, its cells become even more active. In addition to simply dividing, they also begin to contract and move.

A group of cells, contracting in unison, can change the entire shape of the embryo, as when cells on the embryo's back fold up to form the neural tube, the channel that will house the spinal cord. Other groups of cells detach themselves from their neighbors and move to new locations. Time-lapse movies of developing embryos reveal sheets of cells streaming past each other in simultaneous migration. Cells destined to form the gut move upward toward the mouth region. Two ridges on either side of the back move toward the center to form the embryonic spinal cord and brain. Amazingly, the migrating cells seem to know just where to go. It appears that they follow a chemical trail, as ants and bacteria do to reach a food source.

Pioneer cells setting out for new regions first lay down a matrix — a tangled ductwork of filaments that give the cells something to cling to. This is a little like a growing hedge first assembling a picket fence, and then using the fence for support as it grows over it.

Head Start

The flat embryo rolls up into a tube (top row) as a wide ridge begins to form around a centerline groove. The ridge then "zips up" (center row) to form the future spinal cord. The tube then curls into a "comma" as head and tail begin to emerge (bottom row).

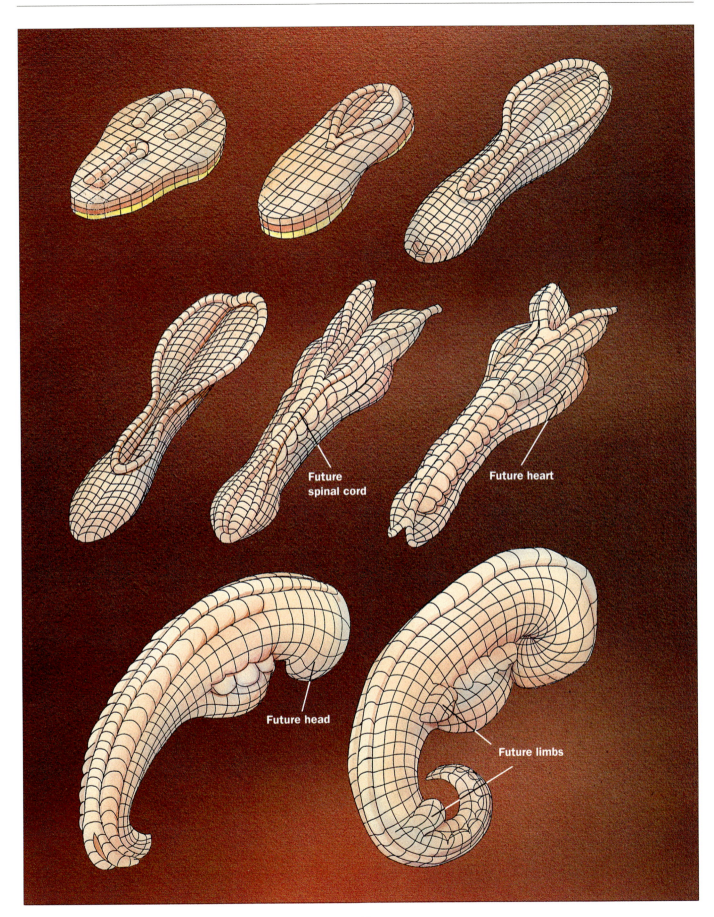

Future
spinal cord

Future heart

Future head

Future limbs

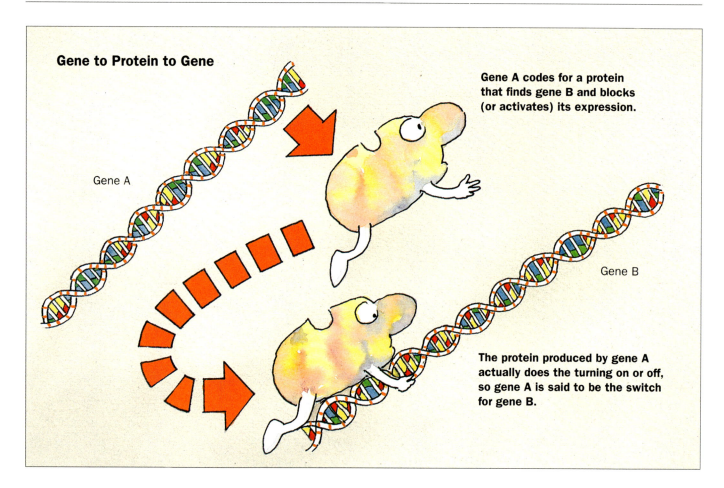

Gene to Protein to Gene

Gene A

Gene B

Gene A codes for a protein that finds gene B and blocks (or activates) its expression.

The protein produced by gene A actually does the turning on or off, so gene A is said to be the switch for gene B.

The Origin of Genetic Circuits

A key evolutionary breakthrough occurred when one gene "learned" how to control another.

HOW ONE GENE CAN TURN ON ANOTHER

To understand how an embryo develops, you need to appreciate how genes act as on/off switches. In Chapter 5, we saw how some genes carry instructions for making worker proteins and other genes code for regulator proteins. Regulator proteins don't make anything or hold anything together. Instead, they go into the nucleus where the DNA is, find a specific gene, and sit on it — thereby blocking it from creating a particular protein. (Some work the opposite way: By sitting on a gene, they signal it to start producing instructions to make a protein. This is what happens, for instance, in the start-up of cell division — see page 158.) In short, regulator genes act as switches, turning on and turning off the genes that code for worker proteins.

When you consider that each individual cell carries all the genes for the entire organism, the necessity of switches becomes clear. The organism wouldn't function properly if, for example, its muscle cells made liver cell proteins. So muscle cells must turn *on* the genes that make the proteins that will turn *off* liver genes. And they keep those genes off for the muscle cells' entire lives. In other words, each cell type — whether muscle, liver, skin, or some other kind, has its own active network of appropriate genes. The rest of its genes — the majority, in fact — lie silent, repressed forever by stubborn regulatory proteins.

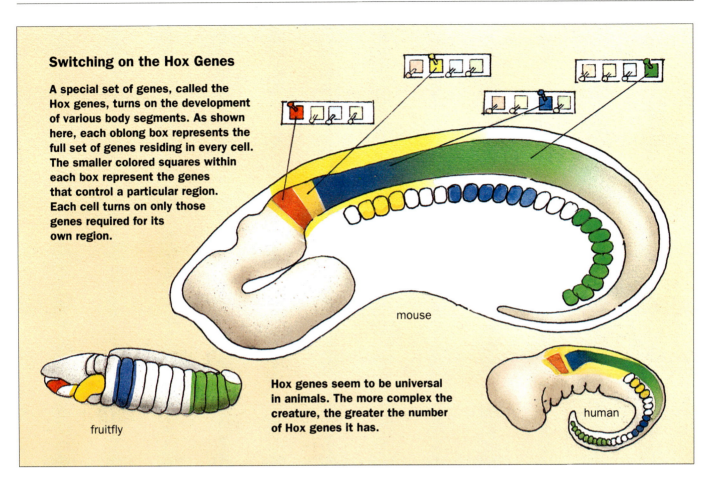

Switching on the Hox Genes

A special set of genes, called the Hox genes, turns on the development of various body segments. As shown here, each oblong box represents the full set of genes residing in every cell. The smaller colored squares within each box represent the genes that control a particular region. Each cell turns on only those genes required for its own region.

mouse

fruitfly

Hox genes seem to be universal in animals. The more complex the creature, the greater the number of Hox genes it has.

human

BODY-SHAPING SWITCHES

Perhaps the most interesting protein switches operate during an embryo's development. As the growing embryo changes shape, each cell needs to know just when to bring the appropriate proteins on line and when to shut them down. Timing is crucial.

If the regulatory genes could talk, they might sound something like this: Gene A: "OK, start making the proteins that define the front end…good, now shut those off." Gene B: "Excellent, now bring in the head-forming proteins," and so forth.

Feedback signals, like those described on page 134, operate the protein switches. Each stage produces the signal molecules that set the next stage in motion. Cascades of worker proteins, governed by a much smaller number of protein switches, swarm into action in wave after wave.

Among the most intriguing of the "master" regulating genes discovered so far are those known as the Hox genes. Becoming active early in the embryo's development, they tell its cells where its head, chest, and lower body should be — and, consequently, where its eyes, arms, legs, etc., will go. If you have ever doubted your kinship with the rest of the world's creatures, you should know that these body-shaping Hox genes are found in insects, worms, fish, frogs, chickens, and cows as well as in humans.

Switches Build on Switches

A single master switch can control a number of subordinate switches. This feature simplifies control over complex operations.

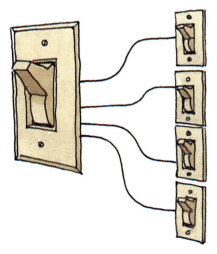

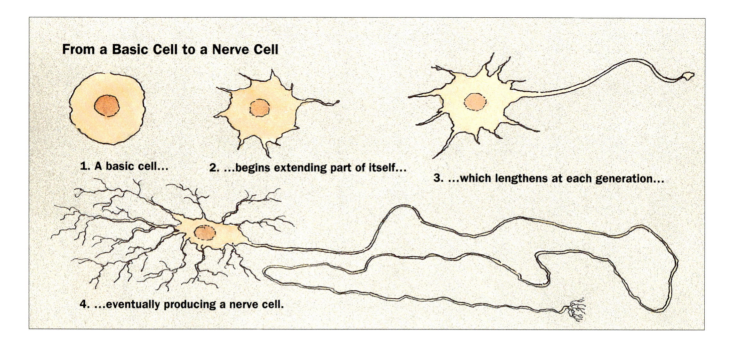

From a Basic Cell to a Nerve Cell

1. A basic cell...

2. ...begins extending part of itself...

3. ...which lengthens at each generation...

4. ...eventually producing a nerve cell.

A COMMUNITY OF SPECIALISTS

We've seen how the embryo progresses from a hollow ball of cells to the beginnings of a recognizable body. Organization proceeds from the general to the specific. First, a blob acquires a top, bottom, front, and back. Next, it develops bands, or rows of cells that define specific regions of the body, and then bumps, which will become head, tail, and limbs. In this process, the embryo's cells polarize, contract, and migrate. They also *differentiate*, taking different paths toward specialization.

The complexity of the tasks facing even the simplest multi-cellular organism demands specialization. This, we might say, is the cell's trade-off — its way of contributing to the whole organism in exchange for food and shelter. Cells carrying within them the capacity to become the whole organism "decide" to become just a small working part of that organism. They do this gradually. During embryonic development, each new generation becomes a little different from the generation before.

At first, the changes are too minuscule to see; then, after a few generations, they're obvious. A small group of cells in an embryo will look like all its neighbors, but as those cells multiply, they become progressively longer and thinner than nearby cells. Within them appear increasing numbers of long stringy proteins that can shorten and lengthen themselves, contracting and relaxing the whole cell. These cells are becoming *muscle*.

Not only must cells do their special jobs in the organism, they must, of course, do them in the proper locations and with the help of the right neighbors. Muscle cells move to a site where a limb will emerge, and find there neighbor cells that will become bone, nerves, and blood vessels.

Skin and Nerve

Nerve cells at first seem no different from their companion skin cells. Yet, as they and their descendants grow and divide, their destiny becomes progressively and irreversibly determined, even though their outward appearance seems unchanged. Then, suddenly, they stop dividing and change becomes obvious: They begin to push out long extensions of themselves — axons — which snake off to make connections with other nerve cells, setting up the brain's wiring circuitry. Thereafter they live as long as their host and their progenitors die off.

In contrast, other members of the original cell population will become real skin cells. Unlike their sister nerve cells that have a single "birthday," lost parents, and a long life, skin cells serve a short time, die, and are continually replaced from persistent progenitors.

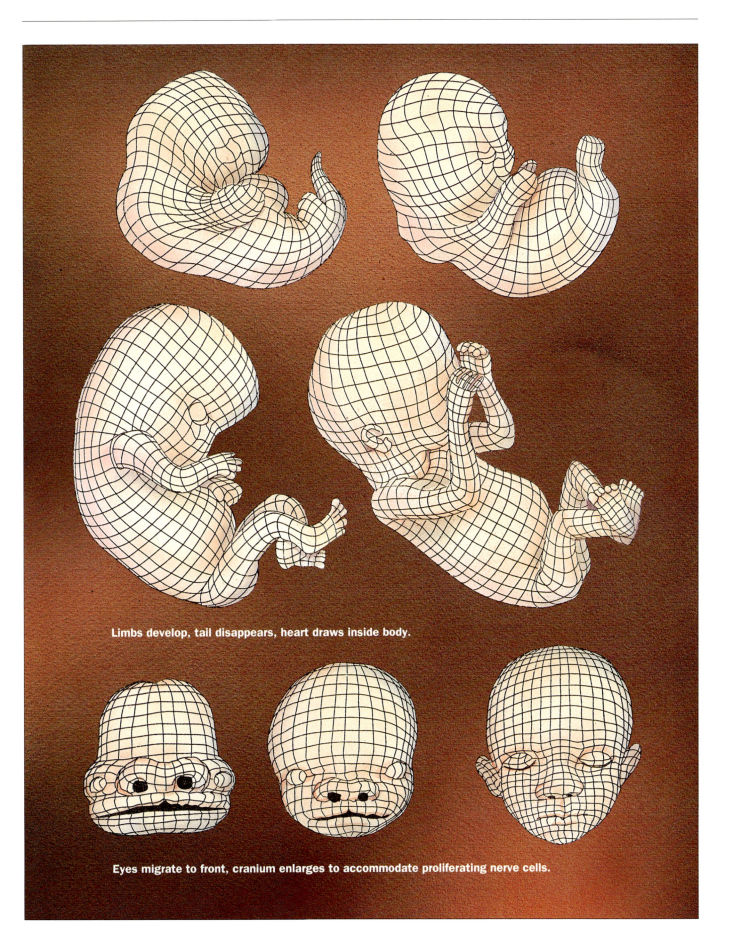

Limbs develop, tail disappears, heart draws inside body.

Eyes migrate to front, cranium enlarges to accommodate proliferating nerve cells.

How Cells Differentiate

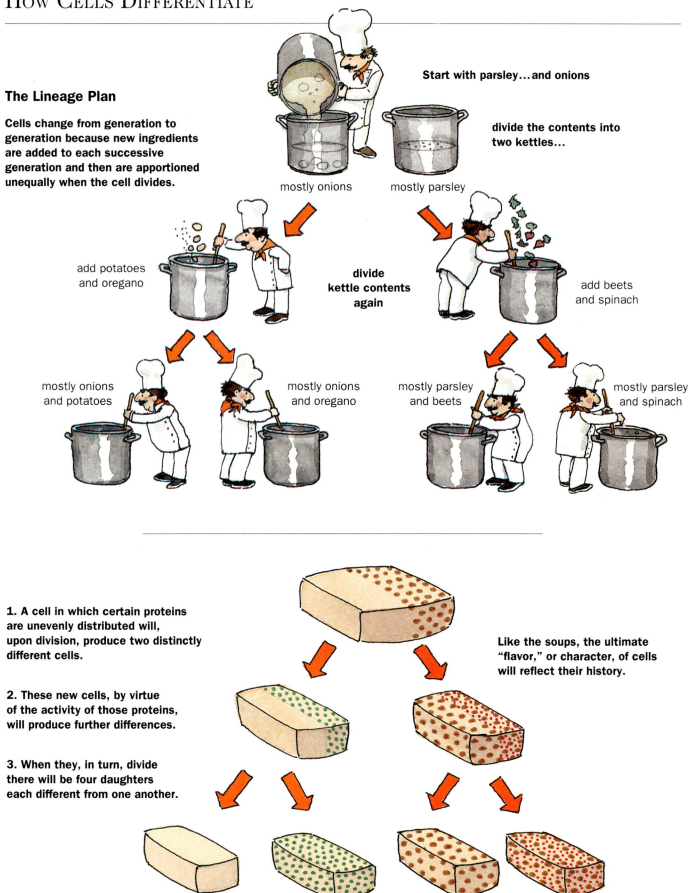

The Lineage Plan

Cells change from generation to generation because new ingredients are added to each successive generation and then are apportioned unequally when the cell divides.

Start with parsley...and onions

divide the contents into two kettles...

mostly onions

mostly parsley

add potatoes and oregano

divide kettle contents again

add beets and spinach

mostly onions and potatoes

mostly onions and oregano

mostly parsley and beets

mostly parsley and spinach

1. A cell in which certain proteins are unevenly distributed will, upon division, produce two distinctly different cells.

2. These new cells, by virtue of the activity of those proteins, will produce further differences.

3. When they, in turn, divide there will be four daughters each different from one another.

Like the soups, the ultimate "flavor," or character, of cells will reflect their history.

THE LINEAGE PLAN AND THE CONTACT PLAN

If every cell in your body comes from a single fertilized egg, how do you end up with so many *kinds* of cells (humans have about 350)? Apparently, this happens in either of two ways: A cell can progressively change itself over generations (the lineage plan), or it can be told to change by a neighboring cell (the contact plan).

The lineage idea is illustrated on the left with a chef and soup model. The chef starts with a simple broth containing one heavy ingredient (onions), which sinks to the bottom of the pot, and one light ingredient (parsley), which floats. When he pours this soup, without stirring it, into separate kettles, the contents of the two new kettles will be different: each will have a different proportion of onions and parsley.

Next, the chef adds more light and heavy ingredients to each of the two kettles. Then he subdivides the contents of the kettles as before, producing 4 completely different soups. You can see how each "generation" of soup is a little different from that before it, yet there's always a little of the original broth in every kettle. Thus the final state of a soup — or a cell — is determined by its history.

At the lower left we show how this principle works with cells. Instead of vegetables, different proteins are coming on line in only half of the cell so that when it divides the proteins are partitioned unequally. This leads to different cells types.

The contact plan, whereby a cell signals its immediate neighbor to change, is simpler. It works by a process called *induction*: One cell sends a message to its next-door neighbor, instructing it to make a particular protein (or proteins). The neighbor responds by making that protein, but only in the part of it nearest the signaler. As shown below, when this neighbor cell divides, the two daughter cells will be different because they will have different amounts of the protein. Repetition of this process over many generations results in an assortment of new and different cells.

The Contact Plan

Cells signal their immediate neighbors to change.

1. One cell induces its neighbor to make proteins in the region nearest it. This makes the neighbor asymmetrical.

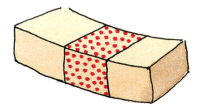

2. The neighbor divides, creating two different daughter cells.

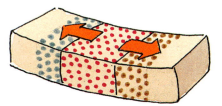

3. The altered daughter cell — the cell in the middle — similarly induces its neighbors.

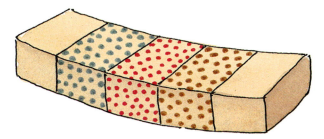

4. When these divide, they, in turn, induce their neighbors to create different daughter cells — and so on.

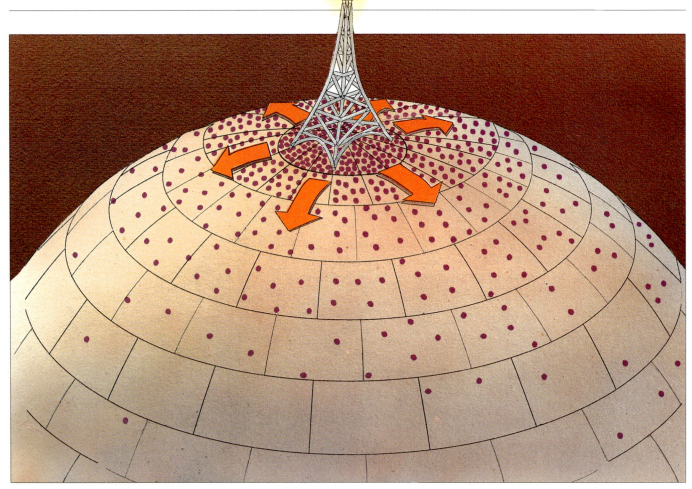

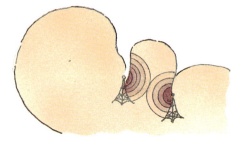

Morphogen gradients develop a limb bud from the trunk of an embryo.

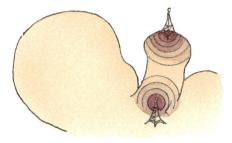

Signal gradients, which originate from key points, command the cells to form the top, bottom, and near and far ends of the limb.

WHEN WHAT YOU DO DEPENDS ON WHERE YOU ARE

In studying how primordial buds of tissue in embryos are sculpted into recognizable body parts, scientists have come to appreciate a special class of molecules called morphogens (the word means "makers of shape"). Morphogens, which are usually proteins, do not act locally in cell-to-cell contacts; instead, they affect all cells over an area of perhaps a square millimeter or two. As their concentration varies, so does their effect on cells.

Imagine a radio tower broadcasting from its position in a single cell. The message received by neighboring cells depends on their distance from the transmitter. Developing cells anywhere within range of the broadcast will read the signal differently, depending on its strength. Nearby cells, since they receive a stronger signal, act in one way; cells further away, since they receive a weaker signal, respond differently, and so on. Beyond the tower's range there is no response.

The resulting morphogen gradients display an impressive versatility. A single family of genes makes morphogens that direct the development of limbs, sex organs, and the brain.

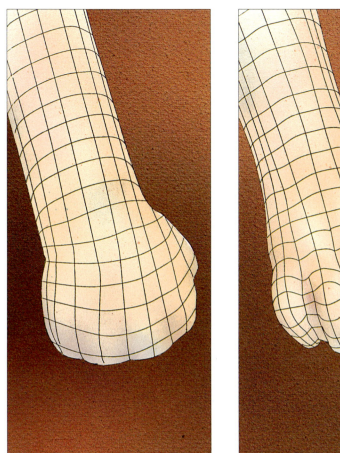

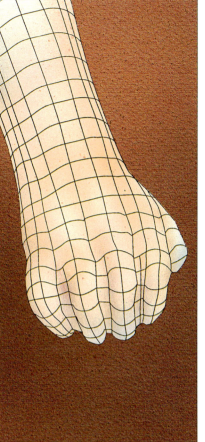

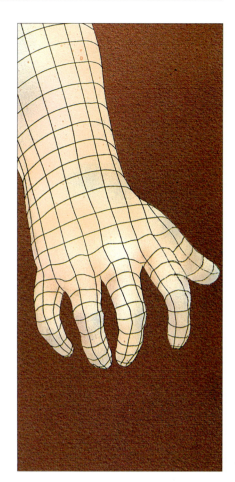

DYING TO CONTRIBUTE

We think of death as the end of life. But cell death plays an essential role in the creation of living bodies — and suicide is part of the program.

While the brain is being built, neurons are produced in profusion — far more of them than the brain will ever use. This overproduction of brain cells, called "blooming," is followed by a massive "pruning" in adolescence. Those brain cells that have made only weak connections to other neurons — or no connections at all — simply die. Some parts of the nervous system lose 85 percent of their neurons in this way! But don't worry; the many billions that are left have been organized and connected by your early experience and are more than ample to see you through adult life.

Your hands also owe their final form to the programmed death of cells. Cells are signaled to create not five fingers but four gaps — the spaces between the fingers. Cells residing in these gaps take their own lives to free up the fingers in the embryo.

BLOOMING: The brain overproduces neurons until adolescence.

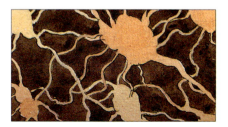

PRUNING: Unconnected neurons commit suicide.

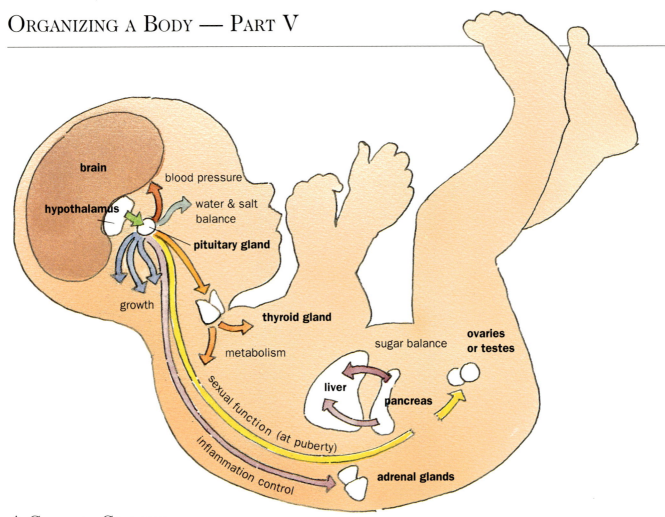

brain

blood pressure

hypothalamus

water & salt balance

pituitary gland

growth

thyroid gland

metabolism

ovaries or testes

sugar balance

liver

pancreas

sexual function (at puberty)

inflammation control

adrenal glands

A Chain of Command

An embryo's development is shepherded by an increasing hierarchy of commands. In the early stages, cells are guided by local interactions; chemical signals pass between only a few cells at a time. Different regions of the body develop semi-independently. Gradually, central channels of communication develop. As cells become more specialized, they become dependent on other cells to do things for them. Cells packed close together in tissues can no longer snatch their food and building materials from their environment the way independent bacteria do. Tissue cells need an elaborate network of blood vessels to bring things to them. They act increasingly in obedience to distant signals traveling in the bloodstream or on nerve pathways.

For example, the thyroid gland makes a hormone that enters the blood and is picked up by cell receptors on target cells causing their metabolism to speed up. Nerve cells develop long extensions of themselves so they can reach muscle cells and stimulate them to contract or relax. The brain gradually takes over as master controller of the nerves and glands, first regulating automatic functions such as heartbeat and blood pressure, then becoming sensitive to sensory signals such as sounds and feelings.

The birth of a newborn infant completes a profound emergence: A quarter of a trillion individual cells, following their local rules, have become one unique being.

More Signals

Hormones, secreted by specialized groups of cells — glands — located in different parts of the body, carry instructions to all the cells through the blood. A master gland, the pituitary, receives signals from the brain and, in response, sends out its own hormones which switch on hormone production in the other glands. As the blood levels of these hormones rise, they signal the pituitary to reduce its hormone output, thereby lowering, by negative feedback, their own level.

Additional hormone control systems maintain other vital balances in the body. For example, the pancreas secretes two hormones, glucagon and insulin, which signal the neighboring liver to make available, or restrict, the supply of sugar in the blood.

Already developing before birth, the sensory organs will enable the organism to interpret and respond to the environment.

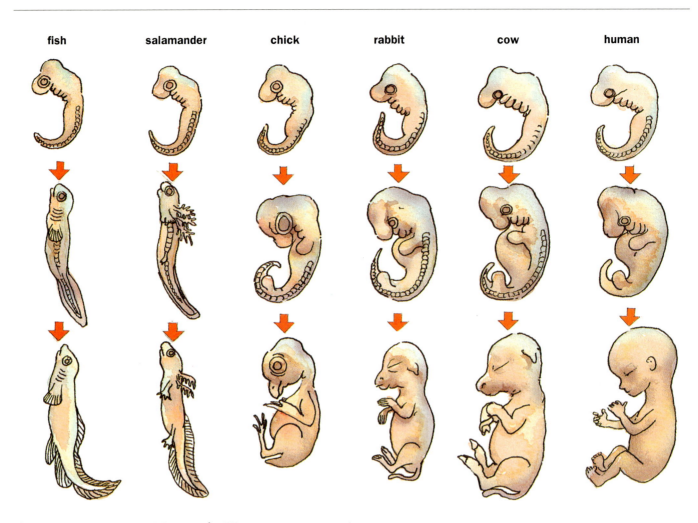

| fish | salamander | chick | rabbit | cow | human |

ANOTHER LOOK AT NATURE'S UNITY

Years ago, scientists noticed with surprise that widely different organisms appear remarkably similar during the early stages of their development. Experts in embryology couldn't distinguish, for instance, between the early embryos of a bird and a human. Why should embryos so closely resemble each other?

This question has led to an appreciation of evolution as a master tinkerer. A tinkerer does not start from scratch each time, but instead makes use of suitable old bits and pieces to build new things. Once a program works well for building a fish, keep it and use it later as a foundation for building a person. It's easier to get rid of a tail and gills than it is to devise a whole new construction scheme.

Over and over in our study of genes and proteins, we discover the effects of tinkering during evolution. The Hox genes of fruitflies are very similar in nucleotide sequence to the genes that perform the same body-shaping function in many other animals, although each species has its own unique sequence. Clearly, these varied genes evolved from a single gene in a remote common ancestor.

At the right are some other examples of tinkering — structures that originally served one purpose and were later co-opted for a completely different purpose.

Embryo Similarity

In the early stages, embryos of most vertebrates are indistinguishable.

174

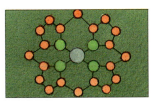
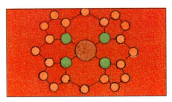

Pigment Molecules — Chlorophyll and Hemoglobin

The molecule that captures sunlight in plants is remarkably similar to the molecule that carries oxygen in an animal's bloodstream.

Chemotaxis — Bacteria and White Blood Cells

Like bacteria drawn to food signals, our white blood cells are attracted by chemotaxis to sites of inflammation.

Development Genes — Fruitfly and Human

The Hox genes that control the basic body layout of the fruitfly are similar to those that shape our bodies.

Proteins — Cheese and Eye

An enzyme in bacteria that is valuable in the manufacture of cheese is also used as the lens of our eye.

Enzymes — Digestion and Blood Clotting

The final folded shape of certain digestive enzymes is nearly identical to that of proteins involved in blood clotting.

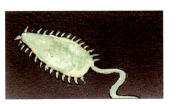

Microtubules — Protozoa and Nerve Cells

Microtubules, the whip-like bundles that single-cell creatures use to propel themselves, also serve as transport "railroads" inside our nerve cells.

Chapter 7

EVOLUTION

The Pattern for Creation

From an evolutionary viewpoint, life is a river of information. Information arises, branches into countless tributaries, and pools in endless combinations. It flows, generation to generation, through the bodies of living creatures, shaping and organizing them along the way. The success of each living thing determines the future of the information it carries. The information is sorted and sifted, with the most useful preserved as it moves downstream. This flow is evolution.

Evolution's principal mechanism — natural selection — is not one process but two: chance and selection. Operating in tandem, chance creates random changes in the information (gene) pool of a population, while selection non-randomly keeps what "works" (that is, what contributes to survival and the production of offspring) and eliminates what doesn't. Nature generates changes in information; changes in information alter life forms; life forms interact with their environment; the environment selects the changes most likely to help the life form survive. Thus, successful changes persevere and are improved upon, which explains why the creatures around us seem so remarkably adapted to their environments. They, and we, are the success stories — at least so far: of all life forms that ever existed, over 99 percent have died out!

Chance and selection are fundamental to any creative act. Chance generates novelty — the new, the unexpected. Selection chooses only those innovations that fit existing conditions. Operating together, chance and selection can produce results so remarkably well-suited to the environment that they give every appearance of having been designed in advance. But while evolution may drift toward greater complexity, it does not — cannot — have a preplanned goal. It simply happens.

An Ancient Earth

James Hutton (1726-1797) pioneered the science of geology. He hypothesized that the earth is very much older than the 6,000 years allotted by Christian dogma. Moreover, it is regularly subject to *slow*, not catastrophic, erosion and sedimentation, as well as periodic earthquakes and volcanic upheaval — changes similar to those we see happening *now*.

Life Has Evolved from Simplicity to Complexity

Jean-Baptiste Lamarck (1744-1829) theorized that living creatures possessed a built-in drive to become increasingly complex — their efforts culminating in humans.

Familiar-Looking Fossils

Fossils often resemble present-day creatures. New life forms don't arise from thin air; there must be connection, progression, transformation.

Selective Breeding

Animal and plant breeders have demonstrated that life forms are not stable and unchanging. By selecting breeding stock carefully, they can produce change readily.

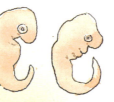

Look-alike Embryos

Embryos of fish, amphibians, reptiles, birds, and mammals — as noted on page 174 — are virtually indistinguishable in their early stages, suggesting that these organisms follow similar developmental patterns and share a common ancestor.

Common Body Plan

Modern life forms share a common body plan. If an organism possesses a rudimentary body part, such as tiny, useless wings on an insect, it suggests that the organism's ancestors had a more useful version of that part.

Similarity of Geographically Isolated Creatures

Creatures living on different continents show related characteristics, suggesting that species migrated long ago and then developed in distinctly different ways.

The Struggle for Existence

Thomas Malthus (1766-1834) suggested that we humans produce more offspring than our food supply can sustain, which raises the possibility that competition for available resources causes creatures to adapt and change.

BEFORE DARWIN

Throughout most of our history, humans have viewed the earth as a static creation of God (or gods), unchanging except for a few worldwide catastrophes like the Biblical flood. Surely the complexity and beauty and fitness of living things meant that they were divinely designed and produced. For centuries, the prevailing world view, first articulated by Aristotle, was that everything had a fixed place in a natural hierarchy — from the most complex Celestial Beings down to the simplest and lowliest of creatures. Fossils were thought to be the remains of earlier creatures produced and then wiped out by God. They had no connection, it seemed, with each other or with living creatures.

The early 1800s was a time of dramatic change: the rise of capitalism, secularism, science, skepticism, and the start of the Industrial Revolution. Our belief in a finite, earth-centered universe eroded as evidence mounted that our planet is a minor player in a universe that reaches far into the unknown. Scientists began to question the assumption that supernatural causes governed natural events.

In rapid succession, a number of discoveries and realizations challenged the older ideas of permanence and divine control.

DARWIN'S INSIGHTS

Many able scientists had noted these developments but no one before Charles Darwin (1809-1882) had put them together into a coherent theory. [Darwin wrote out his basic theory in 1844, but delayed publication out of caution because he saw it as too revolutionary. Alfred Russel Wallace (1823-1913) came up with the same idea independently in the 1850s and wrote Darwin about it. He and Darwin jointly published their theory in 1858.] We may summarize Darwin's ideas as follows:

- Life had a *common beginning*; new forms of life branch off from earlier forms.
- There is *random variation* among individuals in populations and differences continue to arise by chance.
- The pressure of a constantly changing environment in which individuals must compete for survival results in *selection of favorable traits*. Traits that fit well with the environment survive and get passed on to offspring, while traits that do not fit perish.
- While each adaptation is small, *cumulative selection* of favorable traits leads over time to increasingly different forms of life and eventually to new species.

These, together, are evolution.

Further Evidence Supporting Natural Selection

Strengthening the Theory

Darwin's theory has turned out to be one of humankind's most brilliant leaps of the imagination. While Darwin amassed a prodigious body of evidence to support it, it wasn't until the twentieth century that scientists made the discoveries that revealed the hidden mechanisms of the process. We summarize here some of the landmark research.

Genetics

While convinced that the process of selection created diversity, Darwin had no idea *how* life forms actually changed. The physical basis of evolution became clear with increasing knowledge of genetics — the study of the nature of inheritance, sexual recombination, and mutation. (See page 78.)

Natural Selection Observed

Recent studies of finches on a single island show that natural selection can happen very rapidly. In a large population, birds have various beak sizes. When major climatic changes affected the kinds of seeds available as food, birds with better adapted beaks out-reproduced their peers in a single generation. Scientists have observed similar adaptive changes in moth, fruitfly, and bacterial populations. (See page 216.)

Chemical Mechanisms

Study of the nucleotides in DNA and the amino acids in proteins has shown that life's potential to create diversity is far greater than the appearance of organisms would suggest. Furthermore, scientists continue to discover more about how genes get altered, moved around, duplicated, and passed from one organism to another. And, in spite of all its diversity, molecular studies show that a remarkable degree of unity underlies all of life — powerful support for Darwin's view that all life shares a common origin. (See page 122.)

Chemical Relatedness

Scientists can determine how closely related two species are by comparing their anatomy and examining fossil remains. This method has been corroborated by comparing sequences of amino acids in the proteins and nucleotides in the DNA of different species. The more similar the sequences, the more closely related the species. (See page 218.)

Population Genetics

Geneticists view populations of organisms of the same species as "pools" of genes. In the 1930s, scientists began to apply statistical methods to measure numbers of genes in certain populations and how they change over generations. They learned that species of organisms conserve in their gene pools a great capacity for diversity, which makes them enormously adaptable.

Geographic Separation

Naturalists have observed that if a small pool of genes becomes separated from a larger parent pool, for example, if a small flock of birds of one species migrates to an island, the genes in the small pool change relatively rapidly over the generations and eventually become the pool of a new species. (See page 210.)

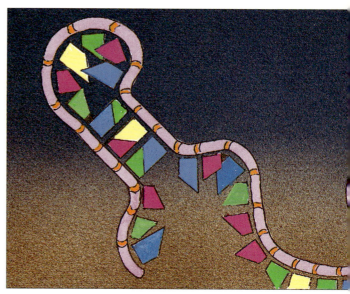

Chains in a Chemical Soup

Before life began, nucleotides were plentiful on earth. Some of them began linking up into chains of RNA, which were able to act as both template (a model for copying) and enzyme (a catalyst to facilitate copying).

Template Meets Enzyme

The RNA chains assumed various shapes depending on the order of nucleotides. Occasionally, two similar chains met up, with one acting as enzyme and using the other as a template.

A Replication Production Line

Linking nucleotides together along the template's length, the enzyme created a complementary copy of the template (and thus of itself). Subsequent copying of the copy led to a chain identical to the first one — i.e., a replicated enzyme. Over time, millions of copies of the original enzyme would be churned out.

Mistakes Create Diversity

These "replicators" inevitably made copying mistakes which, when copied, resulted in differing RNA chains. Some of these variants were improvements on the original; others weren't. Those better able to compete for nucleotides multiplied more rapidly and became the dominant species. Efficient copying had established itself as life's means of information propagation and exchange.

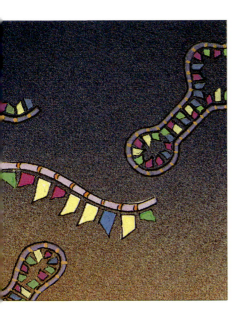

Self-Replicating Chains

Evolution, broadly speaking, is the self-organizing process, not only of life, but of the universe itself. The ordering of matter into elementary particles and then into planets and stars was a necessary prelude to life on earth. Life arose out of the conditions set by those earlier events. Of course, there exists no fossil record before the emergence of cells, but we can make reasonable guesses about how life began.

Our story begins on the steaming, turbulent surface of early earth, some 4 billion years ago, in locations like the hot springs that exist today, where we find bacteria called Archaea. These organisms are known to be very ancient, and they thrive in temperatures near the boiling point of water. Nucleotides and amino acids were probably plentiful before life appeared. Not only can these essential building blocks for DNA, RNA, and protein be made surprisingly easily and thus may have been assembled spontaneously right here on earth, but they have been found in space dust and meteorites that are likely to have showered down on earth over long stretches of time.

Long chains of phosphate molecules, called Poly P, found today in volcanic condensates and oceanic steam vents, may have provided the triphosphate ends of the early nucleotides, thereby giving these molecules the energy they needed to bond to other nucleotides. Once the first nucleotide chains — probably RNA — had been formed, some of them developed a remarkable ability: They could copy themselves. They were not alive in any sense, but simply floated about in the "prebiotic" soup, mindlessly self-replicating.

A self-replicating molecule needs at least two special properties: (1) It must be a *template* — a sequence of units (nucleotides) along which a complementary sequence of similar units can be ordered. (2) It must be an *enzyme*, able to pull free nucleotides from the surroundings and bond them together along the template. We now know that RNA, and only RNA, can perform *both* of these functions. Thus, the earliest self-copying system may have been a mix of similar RNA chains, able to perpetuate themselves endlessly.

How could such a rudimentary self-copying system *evolve* into something that could link amino acids together to make proteins and finally surround itself with a membrane to become a living cell? Quite simply by making mistakes. Occasional inevitable errors in copying — nature's typos — produced a *variety* of RNA molecules, some better at copying than others. The faster copiers prospered because they could interact with amino acids and begin to order them so as to produce more effective protein catalysts, transfer RNAs, ribosomes, and other cell parts.

A Brief History of Life

4.5 Billion Years Ago

4 Billion Years Ago

Accumulating Information from Soup to Brains

During most of life's nearly four billion years on earth, tiny unicellular and multi-cellular creatures living in water were hard at work setting the stage for the big, showy creatures that appeared only in the last half billion years of the drama. Frogs, dinosaurs, trees, birds, mammals, and all the rest arose as elaborations of developmental scenarios worked out by players too small to be seen.

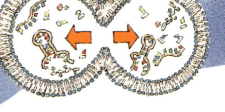

The Earth Cools

As the outermost crust cools, h[...] and gases escape through crack[...] and volcanoes.

A Condensing Cloud of Gas

Gravity compresses the particles in hot gases, forming our planet.

Cell Division

Under pressure from its accumulating contents, a single compartment divides into two.

Proteins

RNA molecules evolve a code for amino acid sequences and begin to assemble crude proteins.

DNA

DNA takes over as the information carrier. RNA becomes the functional link between DNA and amino acids.

Fermentation

Sugar-converting enzymes make limited amounts of ATP, which supplies energy for cells' activities.

Photosynthesis

Some microorganisms "learn" to convert sunlight to sugar, thus tapping an inexhaustible energy source to make food.

(Colors on this time line correspond to the colors on the path of evolution below.)

3 Billion Years Ago

Water and Clay Deposits

Rain and steam create oceans and ponds. Evaporation produces a rich, soupy breeding ground.

Atmosphere

Hydrogen, nitrogen, carbon dioxide, and possibly ammonia and methane hang in the air and dissolve in the water.

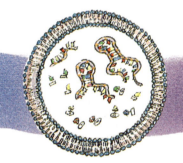

Compartmentalization

Fat molecules spontaneously assemble into bubbles or compartments, sometimes trapping RNA molecules inside.

Self-Replication

Nucleotides begin forming RNA chains. One chain can copy another.

Life's Simple Molecules

Amino acids and nucleotides arrive on space debris or, perhaps, are formed on earth, aided by lightning and ultraviolet light.

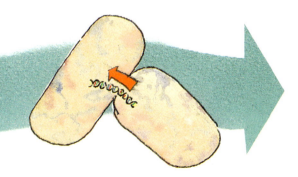

Oxygen Breathing

A few microorganisms "learn" to use the waste oxygen of photosynthesis to make copious amounts of ATP.

Locomotion

Cells develop hair-like cilia and whip-like flagella, allowing them to move around in search of food.

Primitive Sex

One cell injects bits of its DNA into another. New gene combinations proliferate.

Evolutionary Time Line

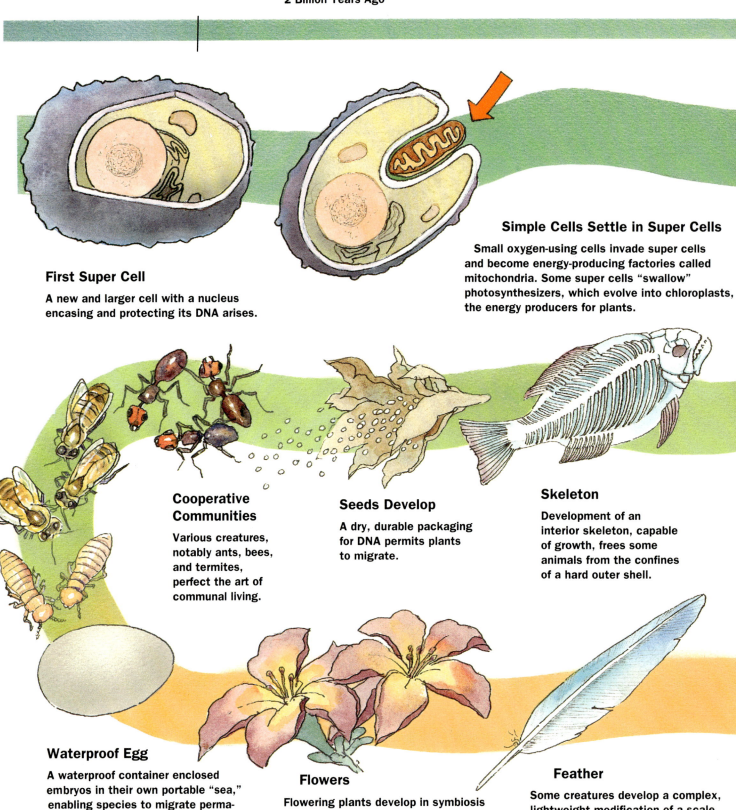

2 Billion Years Ago

First Super Cell

A new and larger cell with a nucleus encasing and protecting its DNA arises.

Simple Cells Settle in Super Cells

Small oxygen-using cells invade super cells and become energy-producing factories called mitochondria. Some super cells "swallow" photosynthesizers, which evolve into chloroplasts, the energy producers for plants.

Cooperative Communities

Various creatures, notably ants, bees, and termites, perfect the art of communal living.

Seeds Develop

A dry, durable packaging for DNA permits plants to migrate.

Skeleton

Development of an interior skeleton, capable of growth, frees some animals from the confines of a hard outer shell.

Waterproof Egg

A waterproof container enclosed embryos in their own portable "sea," enabling species to migrate permanently to land.

Flowers

Flowering plants develop in symbiosis with animals, exchanging nectar for pollen dispersal.

Feather

Some creatures develop a complex, lightweight modification of a scale, providing warmth and, ultimately, the gift of flight.

Multicellularity

Cells begin sticking together in cooperative ventures.

Sex Gets Better

Multicellular organisms produce special germ cells capable of mixing and sharing genetic information.

Body Plans — Animals

Animals evolve radial, then bilateral symmetry (the latter is especially good for locomotion). Segmented bodies allow for complex interaction among parts.

Body Plans — Plants

Plants and animals flourish as they evolve new ways to exploit their environment. Plants tend toward branching tubular construction and radial symmetry.

Central Nervous Systems

Plants and animals develop internal electrochemical signaling. Animal nerve cells will eventually evolve into sensory organs and brains.

Warmbloodedness

Some animals develop higher metabolic rates along with insulation, heat dispersal devices, and internal temperature control.

An Explosion of Innovations

In addition to waterproof eggs and feathers, some warmblooded animals develop binocular vision, opposable thumbs, upright posture, and enlarged brains.

COMBINED INNOVATIONS

Evolution proceeds by gradual tinkering. Complex living creatures had, at one time, cruder and simpler predecessors. Small improvements then accumulated in such a way as to produce big changes over time. Car design offers a useful analogy. For example, the first headlights were dim, removable, oil-fired lanterns; today's models are brilliant, battery-powered, fixed flood beams. As in nature, these changes occurred in small increments, punctuated sometimes by big leaps — as when manufacturers moved the lights from the car's side to the front bumper. At the same time, novelties such as rumble seats and running boards became obsolete and disappeared. Customers' preferences acted as the driving force of selection.

Another feature shared by both creatures and cars: Significant changes often involve the cobbling together of the fruits of several independent and unrelated developments. Production of modern headlights required the invention of the battery, the generator, and plastic "glass," just as development of an eye needed photorecepter cells, an optic nerve, and a transparent lens and cornea.

When comparing car design to creature "design," remember that *evolution proceeds without a foreseen purpose or direction.* Random changes, cumulative selection (i.e., innovations that build on top of prior innovations), and long stretches of time are what allow evolution to work.

An Elephant-Sized Mouse

Life on earth has existed for nearly 4 billion years — such a stupendous stretch of time that it's hard to comprehend its implication for evolutionary change. Here's one example that may help.

Imagine a population of mice that is, for whatever reason, increasing in weight by 0.1 percent each generation. In 12,000 generations, the mice will be as big as elephants. If we call a generation about 5 years — a figure between the actual spans for a mouse and for an elephant — this 100,000-fold increase in body weight will take 60,000 years. This is a very brief period on the evolutionary time scale: if life's almost 4 billion years are seen as comparable to the human life span of 80 years, 60,000 years is about the same as five hours.

The Evolution of Headlights

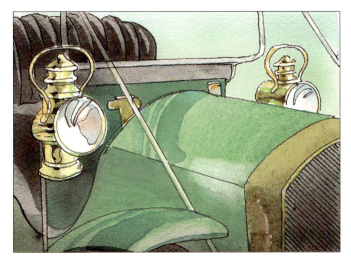

Detachable lanterns hang alongside the driver's seat.

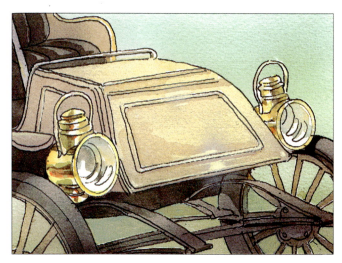

Lanterns move down front to better show the road ahead.

Electrified headlamps are powered by the car.

Headlights are mounted on fenders.

Headlights are embedded in fenders.

Headlights become an integral part of the front end.

Monkeys and Word Processors

Could a roomful of monkeys randomly pecking at their type-writers eventually write a Shakespearean sonnet? This question has been used to challenge the idea that life arose by chance. The odds that a hundred monkeys typing away for a million years could accidentally produce such a work of art are vanishingly small. But if we impose some rules of evolution on the process, we can see how nature increases the chance of success — indeed, how it makes success inevitable. First, let's stipulate that the monkeys type out not Shakespeare's actual sonnets but original sonnets of comparable complexity. That is, we won't demand a specific outcome — just a general pattern. Let's have the monkeys work with word processors programmed to keep successful results and throw out everything else. This rule is the evolutionary principle of selection. Let's arrange for the work to proceed in progressive stages of increasing complexity — another characteristic of evolution. By combining random typing with cumulative "capturing" of successful results, monkeys can write beautiful poetry!

CUMULATIVE SELECTION

1.
First Team: Making Words

Whenever a monkey accidentally types a sequence of letters that the computer recognizes as a valid word, it is saved. "Roses" is acceptable. "Rosgbz" is not. Saved words accumulate over time.

2.
Second Team: Making Sentences

Words generated by team 1 are coded and put into the computers of team 2. When the monkeys strike the keys, the words are strung together in random sequences. The computer saves only those sequences with subjects and predicates — i.e., sentences. "Roses are red" is acceptable. "Roses salad bleakly" is not.

3.
Third Team: Making Sonnets

Sentences generated by team 2 are coded into the computers of team 3. These monkeys randomly order the sentences. Only fourteen-line sequences conforming to the sonnet form are saved.

Lovely…

4.
Fourth Team: Publishing Sonnet Collections

Monkeys in team 4 randomly collect team 3's sonnets into groups, which are printed in bound books. Most sonnets would be nonsense, but a few would be coherent. A tiny fraction of a large enough sample might even be beautiful.

Sale on poetry

5.
Books Are Offered To The Public

Only those books that sell out are reprinted. Thus, the worst poetry is "selected out." The best is kept. Eventually, given enough time, a good collection of sonnets will emerge.

The "Ingenuity" of Chance

When we admire a bird in flight and are told that this ingenious being's ancestor was an earthbound lizard, it's pretty hard to swallow. Yet, as demonstrated by our sonnet-writing monkeys, small changes that arise fortuitously can be "saved" and even amplified at each generation by the addition of further advantages. Given enough time, they can produce something that's never been seen before.

A possible scenario for a bird's beginnings is pictured on the opposite page. In any population of individuals, one born with a slight advantage is more likely to grow up and have offspring that will be like it. Evolution's rules are such that even the slightest advantage will take hold, spread, and eventually dominate in a reproducing population.

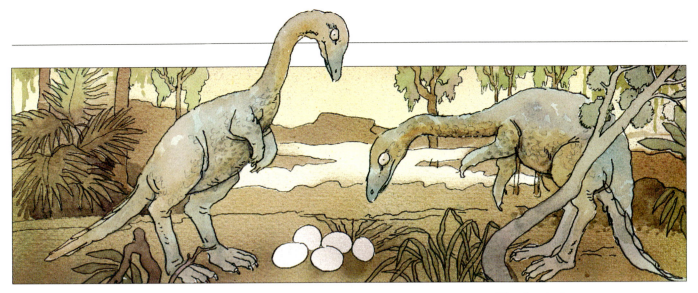

As some reptiles moved toward warmbloodedness, their scales evolved into feathers, providing insulation.

Imagine a baby born with lighter bones than its nestmates — a "freak" produced by a chance variation in bone cells.

When the nest is invaded by a predator, the variant offspring's lighter bones and feather-like scales give it a little extra lift, allowing it to jump away from death's jaws. The double advantage will be passed on to its offspring.

MULTIPLE CHANGES

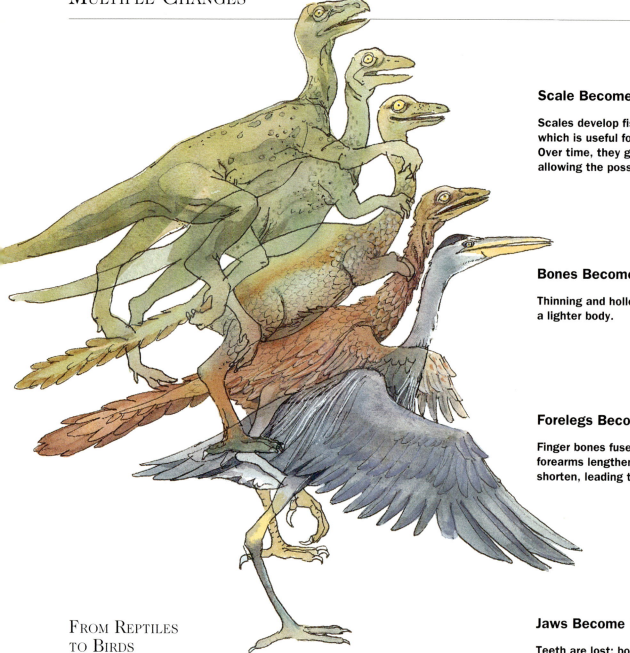

Scale Becomes Feather ▶

Scales develop fissures, trapping air, which is useful for insulation. Over time, they get lighter and longer, allowing the possibility of flight.

Bones Become Lighter ▶

Thinning and hollowing of bones make a lighter body.

Forelegs Become Wings ▶

Finger bones fuse and lengthen; forearms lengthen as upper arms shorten, leading to wings.

FROM REPTILES TO BIRDS

It takes a lot more than feathers and light bones to turn a reptile into a bird. The reptile would need to develop internal navigational equipment, responsive to celestial patterns and the earth's magnetic field. Its eyesight would have to become keener to spot food on the ground. The greater and more sustained energy demands of flight would require major adjustments in its body's ability to maintain a steady temperature. Its reptilian forelegs would have to become aerodynamically efficient as they evolved into wings and its breastbone would have to become keeled to provide leverage for wing muscles.

While these modifications would arise independently, with each small step conferring some advantage (or, at least, doing no harm), they would also reinforce each other in contributing, generation after generation, to the proto-bird's long-term survival.

Jaws Become Bill ▶

Teeth are lost; bone gives way to horny material, elongating into a beak to be used for grasping, preening, and probing.

A Toe Turns Backward ▶

The first of four toes swings backward, making it useful initially as a weapon and later for perching and grasping.

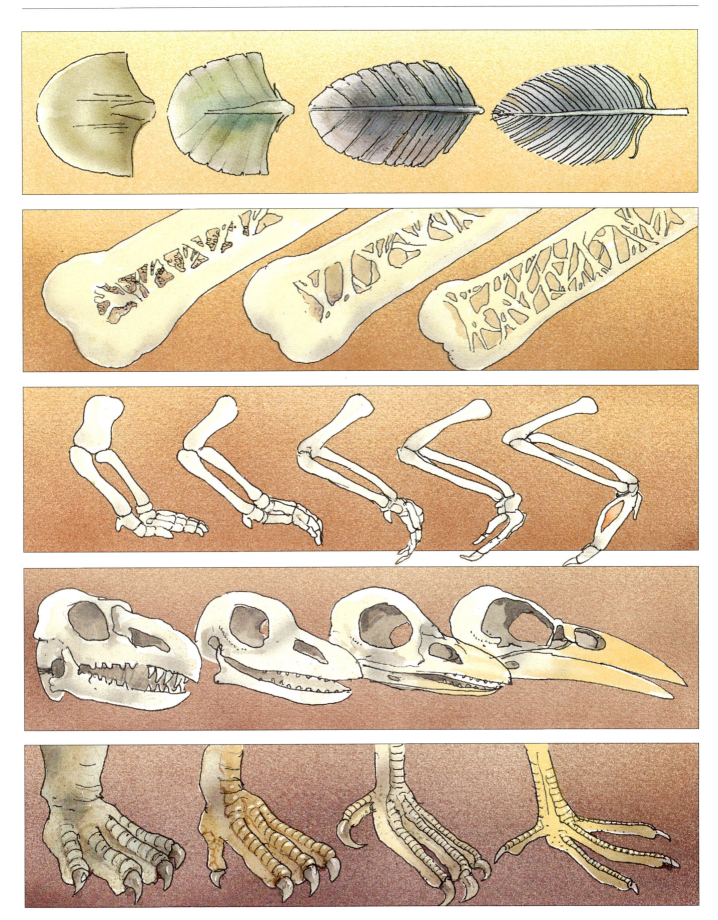

VARIATION AND SELECTION

A HERD OF WILDEBEESTS

Each year in Africa's Serengeti, great herds of wildebeests begin a 600-mile migration. During the journey, some of the animals are eaten by predators, some drown crossing rivers, some die of injury or disease. Some of the deaths are due to bad luck; but, overall, the faster, stronger, and more alert survive the trip, while the less well-endowed are weeded out along the way.

A HERD OF DNA

Now let's shift focus and consider the wildebeests not as a herd of animals but as a vast pool of information. The information exists as separate sets of genes called genomes — each of which resides in an individual animal. And while these information sets are similar — each one of them, after all, spells out "wildebeest" — *each one is also unique.* There could be no evolution without such individual *differences* among genomes. The information in each wildebeest's genome provides the tools — the capabilities of that individual animal — for negotiating the trip. Along the way, some genomes are destroyed. The "best" information sets are those that survive the journey. The genes that are removed by death are, therefore, not just any genes. They are, on average, those that make wildebeests less likely to survive. Individuals die, but the population benefits by an average improvement in its gene pool.

Information Sets

Every cell of every wildebeest contains two sets of genes (simplified here as two short DNA helixes) — one from its mother, one from its father. Every set differs in small ways from every other set, as indicated by the color differences.

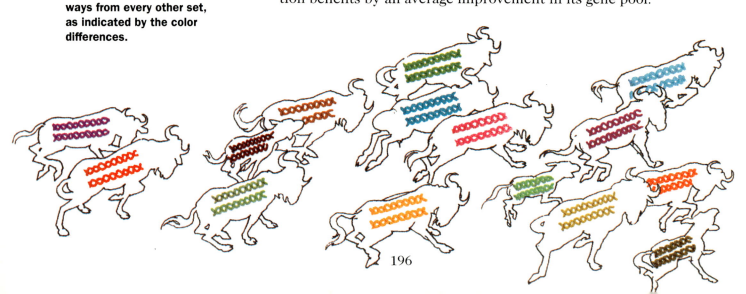

Death on the Journey

Some animals are eaten…

Some drown…

Some succumb to disease.

A Loss of Information

Predation, drowning, and disease remove some information sets from the pool.

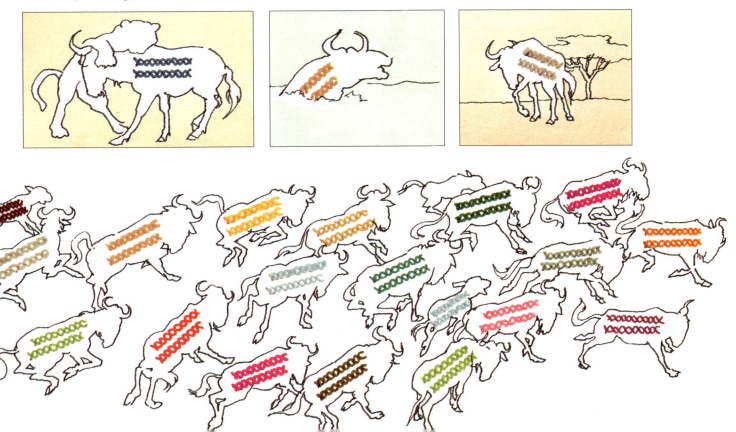

MIXING THE SURVIVOR'S GENES

During migration, the wildebeests mate. A female chooses from among her suitors the male that most impresses her by displays of prowess indicating his dominance over other males — a final selective test among the herd's information sets. Mating is a mingling of sets of genes.

Even before mating, in the ovaries of each female and in the testes of each male, the mother-father pairs of gene sets are first thoroughly mixed and then randomly packaged into eggs and sperms (see page 200). After mating, the fertilized eggs develop into calves. Every cell of a calf contains a new genome, which is likely to be enriched in survival skills and mating success. And all these "better" genes are arranged in new combinations, ripe with the potential for novel characteristics. This is evolution's way of ensuring that any improvements that arise in individuals are distributed as widely as possible to later generations.

Each parent contributes half of its genes to the offspring. Although the parents' individual genes have been successfully "road-tested," the unique combination of genes in their offspring is not yet tried and true.

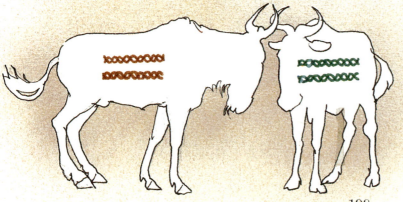

Male and Female — Two Separate Strategies

The making of new life requires an equal contribution of genes from both a male and a female. But the female makes a substantially greater overall contribution to the ultimate success of the endeavor. While a sperm cell is not much more than a set of genes, an egg supplies, in addition to genes, the food, the energy-generating machinery, and the protein-making capability needed to launch the new life.

Furthermore, in many species, the female's body provides the environment for development of the fetus, and, among mammals, the female often continues to nourish the offspring after it has left her body. Owing to her larger reproductive investment, the female is choosier about her mates. Conversely, the lower investment of the male makes him less discriminating and more available. Thus, biology's basic sexual pattern: aggressive males propose; discerning females choose.

She's beautiful!

Hmmm... Strong legs, shiny coat, good teeth...

THE MECHANICS OF GENE MIXING

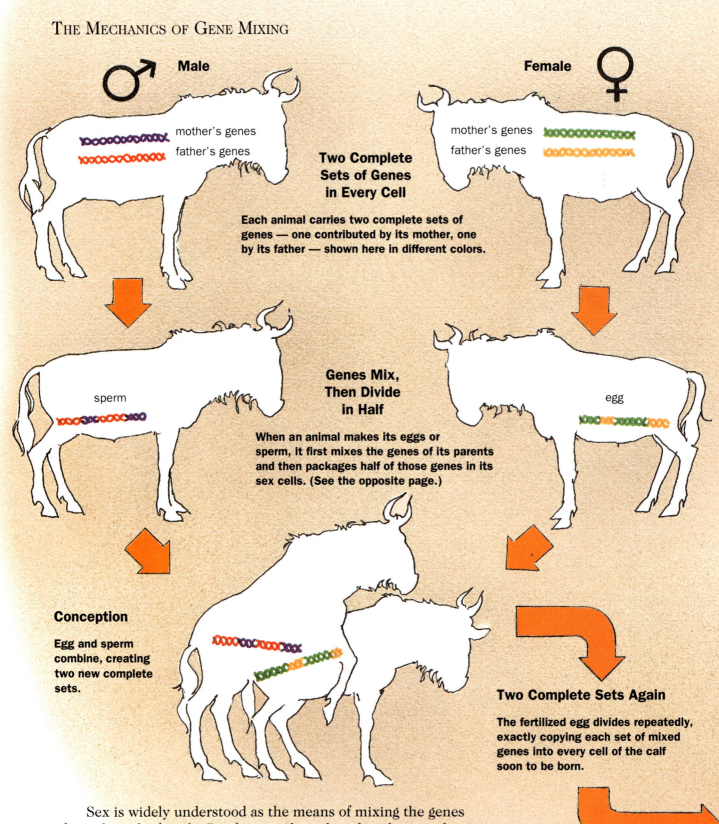

Male

mother's genes
father's genes

Female

mother's genes
father's genes

Two Complete Sets of Genes in Every Cell

Each animal carries two complete sets of genes — one contributed by its mother, one by its father — shown here in different colors.

Genes Mix, Then Divide in Half

sperm

egg

When an animal makes its eggs or sperm, it first mixes the genes of its parents and then packages half of those genes in its sex cells. (See the opposite page.)

Conception

Egg and sperm combine, creating two new complete sets.

Two Complete Sets Again

The fertilized egg divides repeatedly, exactly copying each set of mixed genes into every cell of the calf soon to be born.

Sex is widely understood as the means of mixing the genes of a male and a female. But few people realize that the actual mixing occurs prior to mating in the body of each animal (or plant). Mating is simply the means of bringing together already randomly mixed gene combinations. Here's a simplified picture of how it works.

WHY DO IT?

SEXUAL REPRODUCTION CREATES NEW GENE COMBINATIONS

Since all your cells carry all the information to create a duplicate of you, *any* cell of your body could, theoretically, develop into a perfect copy of you. Many plants form offshoots that detach and become new plants; single cells taken from any part of a plant can grow into a whole new plant. Animal cells can be induced to do a similar thing in the laboratory. For instance, frog skin cells inserted into a frog egg from which all DNA has been removed will develop into a new frog. Thus, the skin cell's DNA contains all the information to make a complete frog. Furthermore, many kinds of multicellular organisms simply "bud" identical offspring from their bodies.

So why then do most creatures reproduce using sperm and eggs? Why do we need sex? It seems wasteful for half a population to produce eggs and half to produce sperm to fertilize those eggs: It means that only half the population can actually produce offspring. Why don't we simply "bud" children directly from our bodies? It seems much more tidy and efficient; and wouldn't it allow us to rapidly out-reproduce sexually reproducing organisms?

Any organism that produces exact copies of itself — whether a single cell dividing or a multicellular organism budding — can't readily adapt to changes in its environment; the only changes its species gene pool will undergo are those caused by mutations. Given this limitation, the species is likely to evolve relatively slowly. Because a sexually reproducing organism has two copies of every gene in its cells, it has a *spare*. (This spare gene can undergo change by mutation and may become useful to a later generation of the organism.)

Furthermore, there is clearly a tremendous evolutionary value in mixing genetic information, in making all sorts of new combinations of already "proven" genes. Some combinations of genes are bound to be winners and lead to new ways of adapting to the changing environment. For example, parasites have a harder time causing trouble if the species they're trying to infect is constantly changing by mixing genes. Sex takes advantage of the favorable attributes of two individuals to create newness.

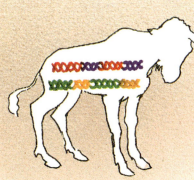

Making Eggs and Sperm

A chromosome is a spool of DNA containing thousands of genes.

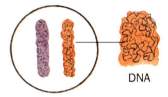

DNA

In special cells in the ovary and testis, chromosome pairs double,...

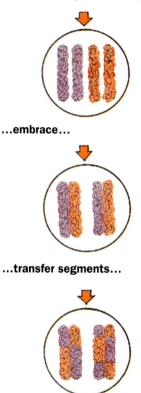

...embrace...

...transfer segments...

...separate into two cells...

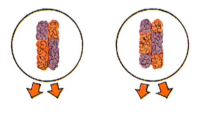

...then into four eggs or sperm...

...each with a different combination of mother's and father's genes.

201

MUTATIONS

HOW CHANCE EVENTS INTRODUCE NOVELTY

Mutations are chance changes in the nucleotides of DNA. When DNA is being doubled during cell division (see page 90), mistakes — like typographical errors — are occasionally made: The wrong nucleotide is inserted into the growing strand. That error will be copied into all future generations of DNA.

Any error in DNA is automatically copied into messenger RNA; so the protein made from such a messenger may have one differing amino acid. Depending on what part of the protein is altered, this change may damage the protein's function, have no effect on it, or, rarely, improve it.

Most mutations do not improve the capabilities of an organism that has already been perfected by evolution over millions of years — any more than random letter substitutions improve a poem. But, once in a while, a mutation does confer an advantage, which will then be passed on to offspring. It is these rare novel improvements in protein function that account for much evolutionary innovation. Randomness introduces newness.

GENETIC TYPOS

A mutation is a mistake which changes the information content of a gene — just as a small typographic change can alter the meaning of a sentence:

A stitch in time saves < $\frac{nine.}{none.}$

He who laughs < $\frac{last}{least}$ > *laughs best.*

Copying Errors

As we've seen (on page 94), DNA is copied with great accuracy...

...but occasionally a wrong nucleotide gets inserted.

Damage to DNA

Radiation (ultraviolet, x-rays, etc.) or toxic chemicals can occasionally damage a nucleotide...

...breaking it so it's "unreadable."

During copying, the wrong nucleotide may get inserted.

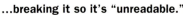

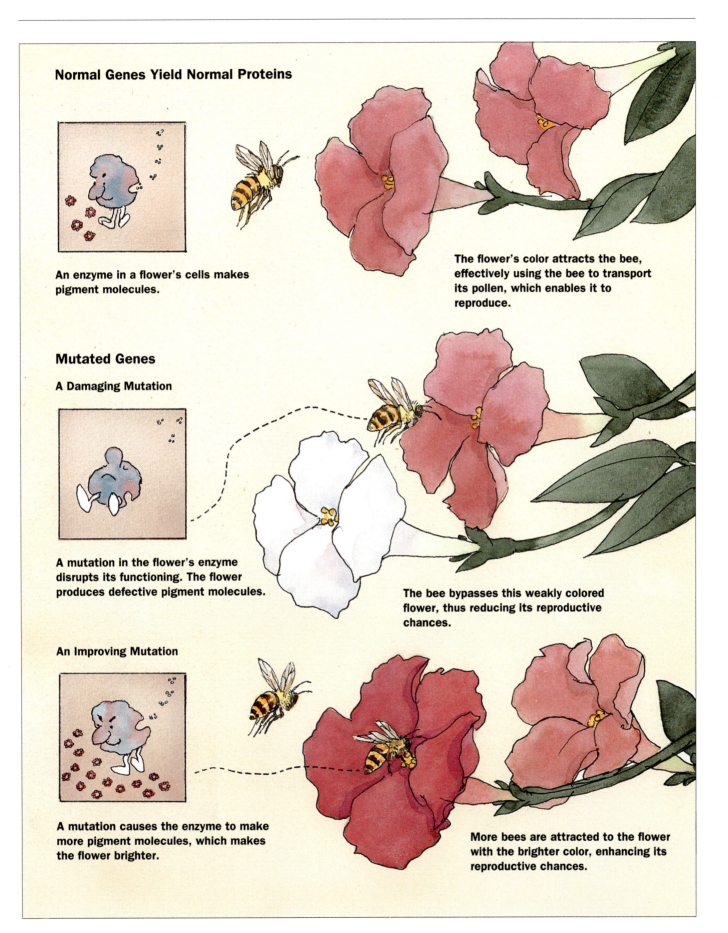

Normal Genes Yield Normal Proteins

An enzyme in a flower's cells makes pigment molecules.

The flower's color attracts the bee, effectively using the bee to transport its pollen, which enables it to reproduce.

Mutated Genes

A Damaging Mutation

A mutation in the flower's enzyme disrupts its functioning. The flower produces defective pigment molecules.

The bee bypasses this weakly colored flower, thus reducing its reproductive chances.

An Improving Mutation

A mutation causes the enzyme to make more pigment molecules, which makes the flower brighter.

More bees are attracted to the flower with the brighter color, enhancing its reproductive chances.

CREATING NEW PATTERNS

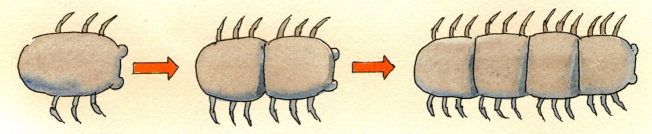

Added Segments

A mutation in a gene controlling the body organization of this hypothetical organism accidentally makes a duplicate body... creating a "Siamese twin" offspring. In future generations, the gene repeats its mistake, adding more body segments.

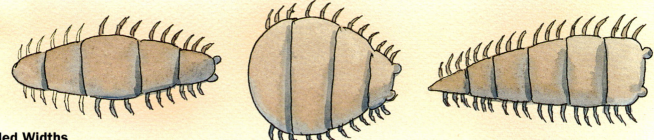

Graded Widths

Mutations in another gene create progressive gradations of segment widths, which cause the organisms to taper or bulge in various ways.

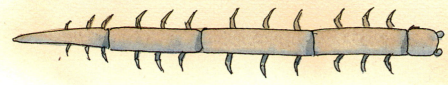

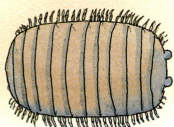

Lengthened or Shortened Segments

Mutations in still another gene lengthen or shorten the segments.

Sometimes mutations in developmental genes give rise to a whole repertoire of body plans.

SMALL MUTATIONS — BIG JUMPS

Ninety-nine percent of the genes of chimpanzees are identical to ours. The remaining one percent of genes somehow give us erect postures and less hair, larger skulls, and bigger brains than chimpanzees. These are almost certainly "switch" genes — ones that turn on or off other genes during embryo development. (See page 164.) Small delays in timing during the formation of our skulls and brains, for instance, could contribute to our larger craniums and greater reasoning ability.

Similarly, consider the difference between giraffe necks and human necks. Both have the same number of bones (seven), but a giraffe neck vertebra can be seven inches long while a human one is less than an inch. Imagine, in the embryo of a primitive giraffe, a heritable gene "defect" in a switch gene in neck bone cells that causes it to be stuck in the "on" position, thus creating longer-than-usual vertebrae.

Specialized Segments

Further mutations allow segments to vary in function — some serve for attaching legs, some for encasing digestive organs, some for reproduction. Segmenting mutations can produce a profusion of body plans.

Certain mutations may be regarded as triggering events — small changes in regulatory proteins that alter when they act, how long they act, their binding ability or other function. Big evolutionary developments may be ushered in at the end of a series of such imperceptible mutations.

Here we illustrate how a a major leap in animal body design could arise from a series of simple changes in segmentation. The "invention" of the segmented body probably started as a mutational mistake that turned a one-part body into a two-parter. The selective success of this fortuitous new layout ensured its rapid spread — and made it more likely that additional segmentations, when they arose by chance, would also be successful.

"Jumping Genes"

Suppose that every now and then someone took one of your books, randomly ripped out a page, and reinserted it in another chapter. Something like this happens in unusual alterations in DNA. Certain enzymes act like scissors, snipping out short lengths of DNA. Other enzymes splice these segments into new locations, similar to what happens during recombination of female and male genes to produce sex cells. (See page 201.) Such *transpositions* happen rarely, but when they do, these "jumping genes" can affect the proper functioning of nearby genes. Scientists aren't sure why these transpositions happen. Just as the torn-out and reinserted page of your book will muddle the text, most transpositions muddle the information in the genes. Occasionally, however, some lead to useful innovations.

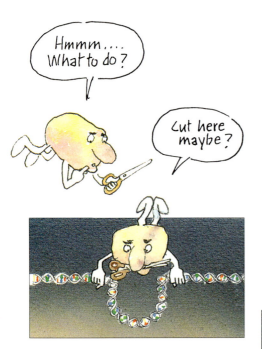

Occasionally a "cutter" enzyme mistakenly grabs a segment of DNA containing one or more genes,...

...and cuts it out of its usual location.

The segment then breaks free, rolls into a circle (a "stitcher" enzyme splices the cut ends together)...

...and moves to a new location on the chromosomes.

CELL NUCLEUS

Certain circular segments of DNA that have split off from the host genome can replicate independently. These are called plasmids.

Corn and Cold Spring Harbor

Remarkably, most of what we know about the molecular mechanisms of inheritance and genetic disease in humans has come from studies in peas, fruitflies, yeast, bacteria, and, yes, corn.

Barbara McClintock, a geneticist at the Cold Spring Harbor Laboratory on Long Island, New York, was the pioneer who discovered jumping genes in the chromosomes of corn plants. Her work revealed that genes are not static but can be rearranged by events that occur naturally inside cells or by outside agents such as x-rays that cause trauma to cells. McClintock's work also led to the realization that there are two kinds of genes: ones that give instructions for specific functions (that is, code for worker proteins) and ones that turn those functions on and off (that is, code for regulator proteins). McClintock was honored with a Nobel Prize in 1982.

PLASMIDS

Sometimes segments of DNA are cut loose by enzymes but are not reinserted elsewhere in the genome. Instead, they roll up into circles and replicate indefinitely as separate genetic units called plasmids, functioning like tiny extra chromosomes.

Plasmids may consist of only a few thousand nucleotides, having just enough genetic information to allow them to replicate themselves independently of their host cell's chromosomes. Or they may have genes that code for proteins useful to the host. For instance, some plasmids in bacteria carry genes that code for proteins that destroy antibiotics — making the bacteria immune to these drugs. Other plasmids enable their hosts to produce poisons that kill other bacteria. Still others make it possible for one bacterium to inject its DNA into another — a kind of primitive sex.

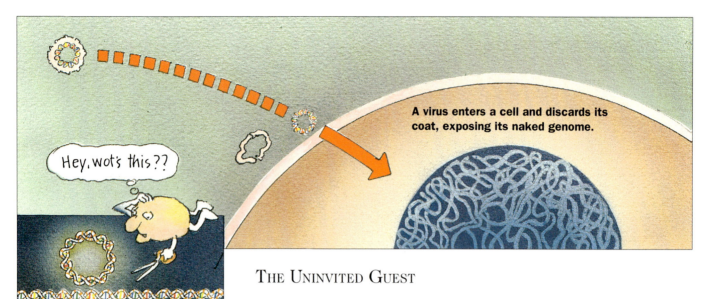

A virus enters a cell and discards its coat, exposing its naked genome.

It approaches a location on the unsuspecting host's genome...

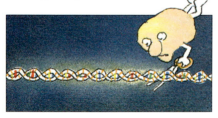

...and gets spliced into it.

When it departs many cell generations later...

...it may take some of the host's genes along with it.

In this way, viruses carry new information from cell to cell.

THE UNINVITED GUEST

Certain independently reproducing pieces of genetic information have become obstreperous over the course of evolutionary history. These plasmids have evolved the ability to use the host cell's ATP and ribosome machinery to make protein coats for themselves — to become viruses. Shielded by their protective coats and making use of certain enzymes they produce, viruses make their escape from the host cell. From there, they invade *other* cells, take over their machinery, and make many more copies of themselves. Viruses can make cells sick (as with the common cold) or destroy their function (as with AIDS, which is caused by HIV, the human immunodeficiency virus). Alternatively, viruses may unobtrusively splice their genes into their victims' DNA, thereby subtly changing the infected cells' genetic character. And when they subsequently cut themselves out of the host's DNA, they may "kidnap" some of the host's genes. Then, as these viruses hop from one infected cell to another, they can transfer normal cell genes along with their own genetic material.

It appears, then, that viruses originally arose from cells and have interacted with cells throughout evolution — sometimes to the cells' detriment when they cause disease, although sometimes in ways that produce evolutionary advantages. Viruses are thus a genetic shuttle between all forms of life.

All this shuffling of genes inside cells and among cells means that life's information is constantly being reorganized. Simple mutations, gene transpositions, sexual recombination, and the existence of plasmids and viruses all contribute to enriching the ocean of variation fished by natural selection.

The Bacterium's Nemesis

A virus injects its DNA into a bacterium.

The invading viral DNA orders the cell's machinery to make multiple copies of it...

...which, in turn, are used to make multiple copies of the virus's proteins...

...which spontaneously assemble...

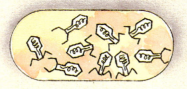

...into new viruses...

...which destroy the bacterium and escape.

Scientists have learned a lot about the relationship between viruses and their hosts by studying the behavior of a peculiar sort of virus called a bacteriophage (literally, "bacterium eater"). Phages (for short) are tiny DNA-filled syringes with protein coats. These marauders go on a take-over mission, attaching themselves to bacterial cells with spider-like "feet" and then injecting their DNA. Information in the phage's DNA prevents the bacterium from using its own protein-making machinery and diverts it to the construction of proteins for the phage. After about 20 minutes, the bacterium is chock full of 100 or so brand-new DNA-filled phages. In the ultimate insult, the phages instruct the bacterium to make an enzyme that breaks open the bacterium's wall. This kills the bacterium and releases the phages to go on to infect other cells!

Occasionally, after a phage has injected its DNA into a bacterium, nothing appears to happen; the bacterium goes right on growing. The phage's DNA, in this case, has spliced itself directly into the bacterium's DNA, where it lies dormant. Many bacterial generations later, the phage's DNA may emerge, subvert the bacterium's protein-making machinery to make new viruses, and then burst out of the bacterium to hunt for the next victim. Sometimes, when this happens, the phage carries along with it some of the bacterium's genes, transferring them to the next bacterium it attacks. Thus, all bacteria seem to be connected by viruses in an immense gene pool in which information is constantly reshuffled.

A virus: "...a piece of bad news wrapped in protein."
— Sir Peter Medawar

How New Species Arise

Necessity Is the Mother of Invention

Over life's history, the number of species — a species being a population of organisms that mates only with its own members and produces similar offspring — has increased into the many millions. The wildebeest's trek across the Serengeti is a vivid illustration of how the basic mechanism of change and selection helps a species adapt and yet maintain its essential character. How does this mechanism work to bring new species into being?

We can think of the first form of life on earth as a trunk from which new life forms have branched, and branched again, and again. Each pair of creatures, or species, has thus left behind it, forever, a common ancestor — just as a tree's branches spring from the trunk that produced them. As species mate and multiply, changing slowly or rapidly, depending on the demands, constraints, and opportunities of their environments, they branch farther out from the mother trunk.

Remember that every species' potential for adapting to its environment — to change — hinges upon hidden capabilities residing in its pool of genes. This pool is stirred by sex, mutations, transpositions, and the other kinds of gene alterations we've discussed in previous pages. These produce changes in protein machinery that control a creature's ability to run or swim faster, to see better, to camouflage itself, to produce a useful digestive enzyme, etc.

Species begin to adapt and change when the environment presents new opportunities or dangers. In response, their hidden skills automatically come into play to help them get food, a mate, a home or to avoid becoming someone else's meal. It is in this sense that the environment *selects*, in fact, makes it *necessary* for organisms to use their genetic potential to survive.

Thus, one species may branch into two, each adapted to two different food sources. They might even adapt to the same food if one eats by day and the other by night, or if one species gets much bigger than the other (think of lions and flies feasting on the same zebra carcass).

The most important single factor in the creation of new species is geographic isolation. If some members of a species happen to get separated from a larger group, ending up in a very different environment — on an island or on the other side of a mountain range, glacier, or body of water — they change much faster over the ensuing generations. If members of this new species were to be reintroduced into the original population, their genetic pool would be so different that they would no longer be able to mate and bear offspring with them.

In this drive toward incredible profusion of form and function, we discern a *life force*: The molecular machinery of cells, fueled by an inexhaustible flow of energy from the sun — and abetted by chance changes in genes and selection of those changes by the environment — inexorably propels life toward greater complexity.

Beak Performance ▶

Isolated on different islands in the Galapagos, finches have, in the course of becoming food-gathering specialists, evolved into different species. All these finches resemble species from the South American mainland more than they resemble finches found in other parts of the world.

High-speed Natural Selection

Recently, scientists banded and observed some 20,000 birds over a period of twenty years in the Galapagos. They noted that within the flock there were finches with big beaks, which worked better for cracking tough spiky seeds; others with smaller beaks, better for tiny seeds. After a severe drought, the spiky-seed plants predominated. Predictably, the birds with bigger beaks enjoyed heartier meals and produced more well-fed offspring. Later, after a protracted wet season, the tiny-seed plants flourished, and the small-beak finches regained dominance.

These changes followed Darwinian principles — at a surprisingly rapid pace. It's not hard to imagine a situation in which a single finch population somehow gets separated into two groups — one inhabiting a "dry" island, the other on a "wet" island. Over the generations, we'd expect the two groups to evolve into distinct "big-beak" and "small-beak" species. This is similar to the scenario Darwin encountered when he observed the finches in the Galapagos islands (see bottom right).

The seed-eating species cracks seeds with its powerful thick beak.

The grub-eating species probes for insects in the bark of trees.

The cone-eating species uses its bill to pull apart cone scales and lifts the seeds out with its tongue.

The mixed-diet species forages on the ground for insects and spiders; it also eats fruits and seeds.

ARMS RACES AND SYMBIOSIS

Organisms don't *try* to evolve. But populations of organisms inevitably change because they *have* to adapt to changing environments. And one of the most important features of an environment is the other creatures in it.

Evolutionary changes in one creature will force changes in the creatures with which it closely interacts. If gazelles get faster, cheetahs will have to become either faster or smarter. If grass gets tougher, horses will evolve stronger teeth. And if humans develop antibiotics, bacteria will develop resistances to those drugs. These relationships can loosely be described as "arms races" that take place on an evolutionary time scale. Since each new development precipitates a counter-development, rarely does one "side" get to declare itself the winner. The process, however, can generate innovations on both sides. Here lies a paradox: Organisms end up in a dance of change that has the effect of keeping their relationships the same.

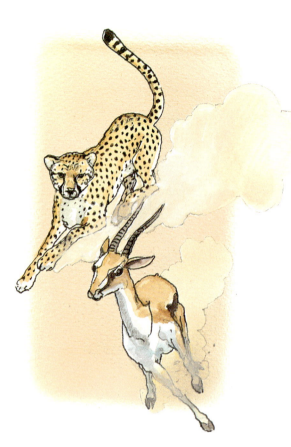

Digesting the Indigestible

Neither cows nor termites can independently digest cellulose — the tough chain of linked sugars that comprises the bulk of grass and wood. Fortunately, both species harbor a particular kind of bacterium in their guts that can do the job. Result: Everyone eats.

cellulose-digesting bacteria

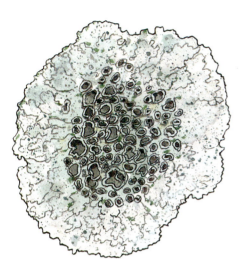

Two Organisms in One

Long ago, certain land-based fungi and water-dwelling photosynthesizing algae found they could expand their mutual horizon enormously by forming a permanent union — they banded together to become *lichens*. The algae provided energy through photosynthesis. The fungi enabled the organism to survive on small amounts of water without drying out. The combo could attach to rocks and other inhospitable surfaces anywhere from deserts to the Arctic — something neither could do alone.

Grazers Perched on Grazers

The oxpecker bird reaps a harvest of ticks and other parasites that live on the hide of rhinos and other large grazers. The birds get a free lunch; the rhinos get free pest control. As an added bonus for the rhino, whenever a predator approaches, the birds swirl upward, loudly warning of danger.

Over time, arms races can mellow out. Enemies settle into cooperative relationships, combining their talents and pooling information. Such a relationship is called symbiosis (see page 30).

A symbiotic leap occurred when bacteria living in the soil invaded the roots of legume plants (clover, alfalfa, and various members of the bean family). The bacteria stimulated the plants' roots to swell, creating nodules in which the invaders took up residence. The bacteria got sugar from the plants, and the plants got essential nitrogen from the bacteria. This was a particularly prized exchange for life in general because, while nitrogen gas is plentiful in air, plants can't use it in that ethereal form. The bacteria, however, can convert nitrogen into earth-bound packages of ammonia and nitrates, which plants *can* use to make their amino acids, nucleotides, etc. Without this symbiotic arrangement, nitrogen could not flow through the living world to sustain multicellular life.

Symbiosis demonstrates the power of combining information in "chunks." Genes from separate species collaborate to produce favorable evolutionary jumps that far exceed the incremental changes produced by hit-or-miss mutations.

CAN HABITS BE INHERITED?

Lamarck vs. Darwin

Jean-Baptiste Lamarck (1744-1829) deserves credit for bringing about a shift in scientific thought about the evolution of life, from judgments based on theological absolutes to inquiries into connections and causes. He conceived the idea that species change over time, and that all organisms are related — essentially the first explicit theory of evolution.

Lamarck is best known for the now-discredited theory of *inheritance of acquired characteristics* — that experiences of an organism could be passed on through inheritance; if an organism worked at something desirable, its children would inherit the fruits of that effort.

Darwin liked Lamarck's ideas about relatedness and change. They fit with his own theory that the accumulation of small changes could accomplish big results. And although Darwin had no explanation for why organisms varied (the science of genetics had not been born) and therefore couldn't rule out the possibility that acquired characteristics were inherited, he felt certain that evolution was not driven by the desires of organisms. Organisms change continuously. Those who coincidentally experience changes that better fit them to their environment have more offspring, and so their kind survives and flourishes.

The perspectives of Lamarck and Darwin differ fundamentally on the matter of purposeful design. Lamarck, while accepting change, couldn't drop the notion of a preordained plan behind evolution. Darwin saw natural selection as a powerful force, lacking a purpose, but creating the illusion of a planned goal.

Darwinism has decisively triumphed over Lamarckism. Evidence accumulated during the last fifty years has firmly established the generalization that information in living systems flows one way: from DNA to RNA to protein; there is no way the environment can influence the organism's proteins to change its DNA.

How Did Giraffes Acquire Long Necks? Two Theories on How Traits Develop

Lamarck's Theory

Back when there were plenty of leaves to eat, giraffes had short necks.

In time, the giraffes stripped the lower branches bare, so that the only available leaves were higher up.

Giraffes had to stretch their necks to reach the higher leaves.

The stretched-neck trait was passed on to their offspring, resulting in long-necked giraffes.

Darwin's Theory

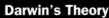

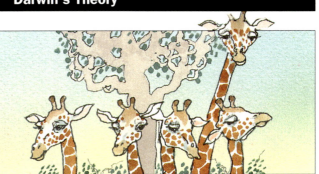

Back when there were plenty of leaves to eat, most giraffes had short necks, but some did have longer necks.

In time, the giraffes stripped the lower branches bare so that the only available leaves were higher up.

The short-necked giraffes began to die off for lack of food; the long-necked giraffes survived and multiplied.

Eventually, only long-necked giraffes were left.

An Experiment in Evolution

As late as the 1940s, scientists did not believe that bacteria, the most abundant and ancient forms of life on earth, were governed by the rules of evolution. Bacteria multiplied and changed so fast that scientists thought their inheritance might be directly changed by their environment (in the Lamarckian mode). Nobel laureate biologist Salvador Luria, however, suspected that bacteria, like giraffes, obeyed Darwinian rules. In 1943, while watching slot machines being played at an alumni dance, Luria conceived an experiment that would conclusively settle the question.

Luria's Question:

Certain kinds of viruses kill bacteria. If bacteria are given food and grown in liquid in a tube for a day, the liquid becomes cloudy as their numbers swell to a billion or so. (Bacterial cells divide about every half-hour.) If you then add the bacteria-killing viruses to the tube, virtually all the bacteria are killed off within 20 minutes. But wait! A day later, the tube again has a billion bacteria in it — all immune to the virus. Is the immunity caused by the virus (the Lamarckian answer) or does an occasional bacterium become immune by chance — whether or not the virus is present — and then multiply to produce a new immune population (the Darwinian answer)?

The Experiment:

Luria put an equal number of virus-sensitive bacteria into each of a hundred tubes and gave them food. The bacteria multiplied for a day. Next, he set out a hundred dishes, each covered with a gelatin-like material containing food *and bacteria-killing viruses*. He then spread the contents of each tube onto each of these dishes so that wherever a cell landed on the dish, it stayed put and began to multiply. A day later, any cell that was immune to the bacteria-killing virus would have multiplied into a clump of cells big enough to see.

The Reasoning:

Luria reasoned that if the bacteria *acquired* immunity — that is, if they somehow "learned" from contact with the virus how to avoid being killed by it — all the dishes would have about the same number of clumps because they'd all have been exposed to the same challenge. If, however, the immunity was caused by random mutations in the bacteria — which would occur whether or not the virus was present — the dishes would end up looking different. Many would have no clumps at all, some would have a few, and a rare one would have many clumps.

Here is Luria's reasoning: Mutations are rare events; they occur once in every 5 million or so cells. If a mutation that made the bacteria immune to the virus occurred *soon* after the cells were placed in one of the hundred tubes, that immune cell would have a long time to multiply and make many offspring; there'd be lots of clumps on the dish — a "jackpot." The *later* a mutation occurred, the fewer the clumps on the dish. Of course, many tubes would produce no mutant at all, so nothing would grow on those dishes.

The Result:

As Luria anticipated, there were widely differing numbers of bacterial clumps on the dishes and no clumps on most. This meant that the mutation that caused immunity was random and uninfluenced by the presence of the virus.

How Are Slot Machines like Bacteria?

How Luria Got the Idea for His Experiment

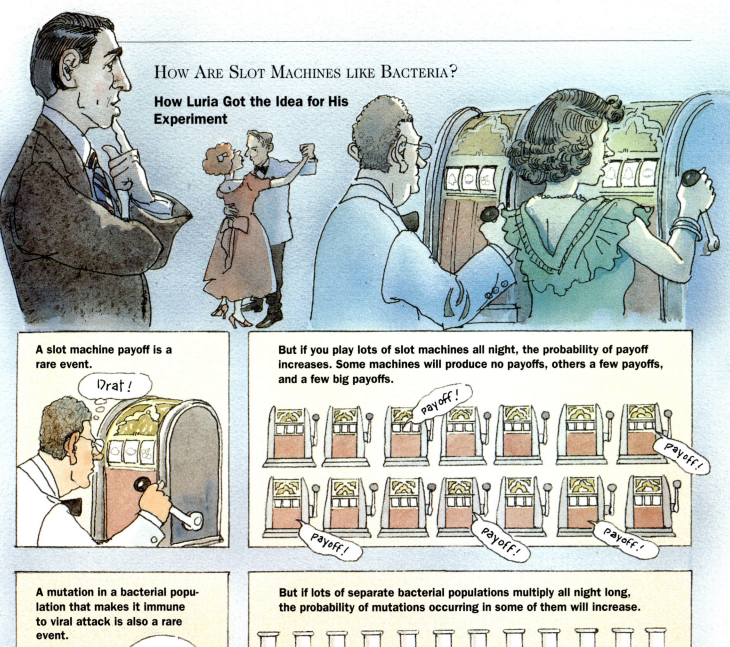

A slot machine payoff is a rare event.

Drat!

But if you play lots of slot machines all night, the probability of payoff increases. Some machines will produce no payoffs, others a few payoffs, and a few big payoffs.

Payoff! Payoff! Payoff! Payoff! Payoff!

A mutation in a bacterial population that makes it immune to viral attack is also a rare event.

Look Harry, a mutant!!

Boy, ya don't see many of those!

But if lots of separate bacterial populations multiply all night long, the probability of mutations occurring in some of them will increase.

mutation! mutation! mutation! mutation! mutation!

It occurred to Luria that it would matter very much *when* such a mutation occurred.

An *early* mutation would produce a huge number of offspring — a jackpot — because the mutant cell's progeny would have all night to multiply. A mutation occurring later would have less time to produce offspring; so there'd be fewer cells in the tube in the morning. This insight suggested an experiment that would conclusively test evolutionary theory.

EVIDENCE OF RELATEDNESS

Comparative Anatomy

These skulls of a human, a gorilla, and an orangutan are obviously related. Do their anatomical features give any clues as to which is more closely related to which?

How Old Is It?

The history of life is written in the earth's rock layers as in the pages of a book. Fossils found in these layers are like the writing on the pages. By dating the rocks, paleontologists can learn the age of the once-living fossils trapped inside them — they can discover *when* creatures lived. The age of a rock can be determined because the naturally occurring uranium atoms in rocks decay into lead atoms at a constant rate over billions of years. The *relative* amounts of uranium and lead in a rock provide a measure of its age.

Radiocarbon Dating

Small amounts of radioactive carbon (carbon-14), generated when cosmic rays bombard nitrogen in the atmosphere, become part of the carbon in all living tissues, as part of carbon dioxide. After a creature dies, the amount of carbon-14 within it steadily decays, emitting radiation. Half of the carbon-14 decays in 5,730 years, half of the remainder decays in the next 5,730 years, and so on. (The time it takes for half of a radioactive element to decay is known as its half-life.)

Measuring the amount of radioactivity left in the carbon of once-living tissue such as bone, skin, or wool can tell us how old that tissue is. (This method is not useful for tissues older than 40,000 years because they don't contain enough radioactivity to measure accurately.)

ALL IN THE FAMILY

We know that all animals and plants are related because they all use the same genetic code and much of the same molecular machinery for living. But how do scientists find out *how* closely any two species are related? That is, when did their common ancestor live? How long ago were they one and the same creature?

Over millions of years, genes accumulate mutations at an overall steady rate. The number of mutations that have accumulated in genes for the same function taken from members of two *different* species is a measure of the species' relatedness: The smaller the differences, the more closely related they are. Assuming that any two species have a common ancestor, the simplest way to connect them is to construct a genealogical tree.

Genealogical Tree

This simple lineage is derived by determining the nucleotide sequence of a particular gene taken from each of three animals: an orangutan, a gorilla, and a human. Out of a sequence of 75 nucleotides in the gene, 12 differed between the human and the gorilla, and 20 between the human and the orangutan. Assuming that mutations have hit these genes randomly and at the same rate over time, this genealogical tree says that humans and gorillas are more closely related to each other than either of them is to the orangutan. In other words, humans and gorillas had a common ancestor that lived more recently than the common ancestor of all three.

218

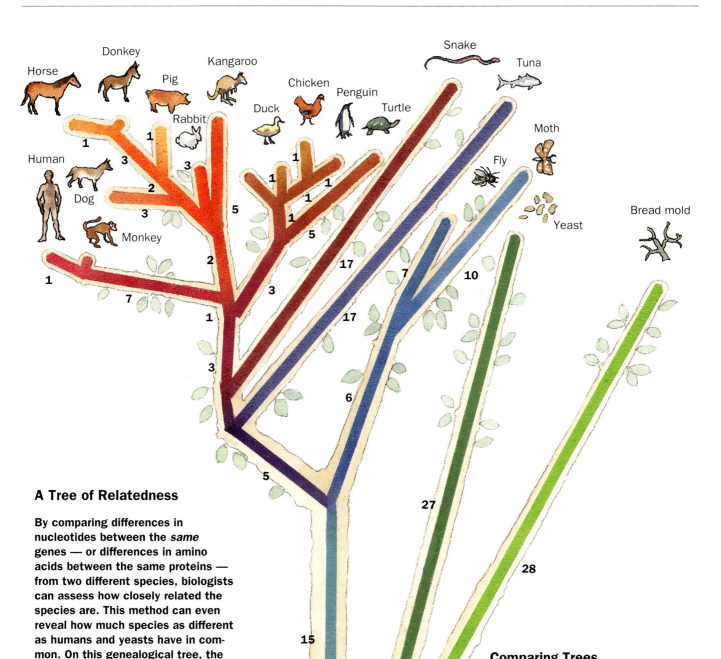

A Tree of Relatedness

By comparing differences in nucleotides between the *same* genes — or differences in amino acids between the same proteins — from two different species, biologists can assess how closely related the species are. This method can even reveal how much species as different as humans and yeasts have in common. On this genealogical tree, the length of each branch — the distance from a common ancestor — has been drawn roughly proportional to the number of nucleotide differences between pairs of species. Thus, for instance, a moth and a tuna fish differ by 38 nucleotides (10 + 6 + 5 + 17); a turtle and a penguin differ by 8 (5 + 1 + 1 + 1); a horse and a pig differ by 5 (1 + 3 + 1).

We can apply this formula for molecular relatedness to all living creatures — as long as they have genes in common.

Comparing Trees

One of the most gratifying experiences for scientists is to have a hypothesis confirmed by two or more entirely different routes of experimentation. When molecular biologists compared their genealogical trees (made by counting differences in the nucleotides of genes) with paleontologists' evolutionary trees (made by dating fossils and by comparing fossil anatomy with the anatomy of living organisms), the trees were remarkably similar. Scientists can combine all these methods to map evolution with ever-increasing detail and accuracy.

Three Brains in One

Our brain has a "layered" architecture, with newer parts built on top of the earlier parts.

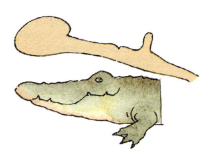

The first and most ancient, the R-complex (the R refers to reptiles), developed as an extension of the upper brain stem. This area influences our territoriality, mating, and aggression — our basic "survival brain."

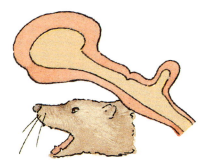

Above the R-complex lies the limbic system (evolved with the earliest mammals), which produces our emotional states — our "feeling brain."

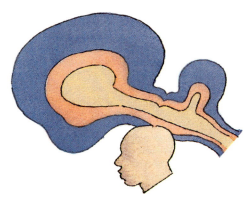

Our cerebral cortex is the thick, outer layer of our brain — our "thinking cap." With this new brain mass, we developed the traits that make us uniquely human.

GENES AND BRAINS

No organ in the history of evolution has grown as fast as the human brain. From the time of our ape-like ancestors, some 5 million years ago, until modern humans appeared about 200,000 years ago, the human brain has added a cubic inch to its total volume every 100,000 years. While adding on all this new mass, our brain's equipment for regulating our bodies and our instincts was left virtually intact; new circuitry was installed right on top of the old (see the diagram at the left).

Humans at some point crossed a threshold; the information content of our brains surpassed that of our genes. There are about 3 billion nucleotides, or bits of information in each person's DNA. Each of us has a brain with some 10 *trillion* units of information, if we define a unit as a single connection between nerves conveying a digital (yes or no, on or off) message.

Our add-on brain was adjustable hardware. It could modify its neural connections in response to experience — a feature essential to learning. Because of the powerful advantage learning confers, the better learners among our ancestors thrived and multiplied. Our earlier brain structures handled survival and reproduction; the add-on systems increasingly opened up the more abstract realms of wonderment and creativity.

Language most probably evolved in stages, perhaps from making simple calls, then to representing objects (that is, naming things), and then to representation of ideas. Our ability to use language to further ideas is a marvelous example of positive feedback. One idea led to another, building new connections in our brains — which, in turn, opened new doors of comprehension. It has been said that not only did we invent language, but language invented us.

The evolution of consciousness, like language, depended greatly upon cooperation. Brains of similar complexity interacted and lifted each other to higher levels of thinking. In this evolutionary phase, which probably occurred within the last 10,000 years, we gained both a sense of self and a sense of time. With "self" consciousness each individual is a "me" — a player in his or her own drama. A conscious person brings a measure of objectivity to an otherwise totally subjective life. With the dawn of consciousness, our species has evolved a mind that can observe itself.

Through the ability to imagine their past, present, and future, our ancestors could look both forward and backward in time. Agriculture, calendars, and a host of other cultural innovations flowed from this insight.

Our ability to look ahead and imagine the way life *could* be has given us a gift with far-reaching consequences: the gift of choice. We can make choices not only in our own lives but, to a degree unimagined by our ancestors, in service of the life of the entire biosphere.

Although the cerebral cortex *looks* symmetrical, its two hemispheres process information differently. Generally, the left side thinks analytically, one-thing-at-a-time, and reductively. The right side thinks spatially, all-at-once, and holistically.

Cultural Evolution

Genes and Ideas

We come now to the furthest reach of evolution's information-building drive: the transmission of ideas through culture.

Ideas, like biological innovations, seem to follow the rules of evolution. From the random stream of thoughts, utterances, and writings, a few get selected and are reproduced, while the rest are weeded out. An idea born in one brain evolves as it passes through other brains. Those ideas that spread best will find more enduring lives in the world's libraries and CD collections.

Any discrete, memorable idea enters the competition: car headlights, microchips, Pinocchio, algebra, natural selection, and the TV jingle that you hate but can't get out of your head. The first order of business for an idea is to spread itself, whether or not it acts for the good of anything. The more it sticks in our collective minds, the more likely it is to survive.

With the rapid spread of ideas, cultural evolution has greatly sped up the rate of change on earth. It has given us the tools to expand our range and lengthen our lives as well as to extract an ever-growing quantity of materials and energy from the biosphere. However, culture must fit within nature. By almost any measure, the natural environment we live in is under stress. Many organisms have been unable to keep pace with the changes we have produced — and extinction rates are rising. Evolution suggests to us that ideas that work for one environment may not work for another. In other words, the ideas that got us here may not be the ones to keep us here.

As some of our most "successful" ideas come to threaten the ecological balance of the natural world, we need to take another look at them and make choices that contribute to the well-being of the life system as a whole. One such choice is what biologist E.O. Wilson calls "biophilia," which he defines as our natural sense of connectedness to life. Wilson writes, "to explore and to affiliate with life is a deep and complicated process in mental development. To an extent still undervalued in philosophy and religion, our existence depends on this propensity, our spirit is woven from it, hope rises on its currents."

It may be that our growing appreciation of that truth will be the greatest legacy we can pass on to future generations.

As we deepen our imprint on the natural world, we increase our responsibility for it.

Illustration after M.C. Escher

NOTES

We have benefited from the study of many books but note here our most valued general source: *Molecular Biology of the Cell* by Bruce Alberts, Dennis Bray, Julian Lewis, Martin Raff, Keith Roberts, James D. Watson, Garland Publishing, 1994. Another useful source was *A Guided Tour of the Living Cell* by Christian de Duve, Scientific American Library, 1984.

Chapter 1. Patterns

p. 2 "Before a single plant or animal appeared..." A good discussion of the contributions made by our microbial ancestors can be found in Lynn Margulis and Dorion Sagan's *Microcosmos*, Summit Books, 1986.

p. 14 "Wrinkly skin provides more surface area than smooth..." Elephants also gain more surface area through their oversize, blood-rich ears — another "creative mistake" of nature.

p. 16 "...like a pair of Mickey Mouse ears..." This analogy came from M. Mitchell Waldrop's *Complexity*, Simon & Schuster, 1992.

p. 21 " The steam engine with a governor..." Gregory Bateson effectively makes the case for life's self-corrective tendencies in *Mind and Nature*, Bantam Books, 1979.

p. 24 " Life Maintains Itself by Turnover." Turnover also occurs in ecosystems through the birth and death of organisms. Individuals come and go but the overall characteristics of populations remain relatively stable.

p. 28 "...from near boiling sulfur springs..." Recently scientists have focused their attention on single-celled microorganisms called thermophiles, which can thrive in deep ocean vents and hot springs at temperatures higher than the boiling point of water. Evidence suggests that organisms like these were the first living creatures on earth.

p. 30 "Creatures are self-interested but not self-destructive..." Appreciation of the cooperative nature of life has been gaining ground in recent years. Lewis Thomas and Lynn Margulis, among others, have written extensively on the evolution of symbiotic arrangements. Another interesting treatment can be found in Robert Axelrod's *The Evolution of Cooperation*, Basic Books, 1994, in which the author uses game theory to show the effectiveness of cooperation as a strategy for survival.

p. 31 "...originally acted as small predators..." The now widely accepted theory that mitochondria were once invading bacteria was championed by Lynn Margulis and later substantiated when it was found that mitochondria have their own DNA, different from that found in the cell nucleus.

Chapter 2. Energy

p. 42 "This constant flow keeps our earth in an energized state..." Visible light, occupying just a tiny band on the electromagnetic spectrum, has just the right amount of energy to bounce electrons into higher orbits — the necessary first step in energy conversion. (Lower-frequency infrared light lacks the strength, while higher-frequency ultraviolet light carries so much energy it tends to break bonds and so disrupt the function of molecules.)

p. 45 "How a Dog Shares His Fleas." This metaphor, taken from Heinz Pagels (*The Cosmic Code*, Bantam Books, 1983) also helps to make clear the statistical nature of the second law of thermodynamics. Individual fleas, like individual atoms or molecules, move about quite randomly. But given another, flealess dog, the probability is that the whole population of fleas will in general flow in one direction — from a more concentrated, or ordered state on one dog to a more dispersed or randomly distributed state on two dogs — until equilibrium is attained. Similarly, atoms and molecules will flow from a more concentrated to a more dispersed state. The probability of their flowing in the opposite direction is vanishingly small. Thus, the unidirectionality of events — time itself — arises from the statistical behavior of the atoms and molecules of matter.

p. 51 "A fourth group, the 'decomposers'..." These organisms complete the cycle of materials by converting the substance of all other life into reusable forms in the soil, whereupon they are taken up once again by plants. All life on earth would quickly cease if the decomposers stopped work.

p. 61 "Making Sugar Out of Thin Air." As mentioned in our note to page 28, there are many kinds of organisms (mostly bacteria) that can construct themselves without the aid of sunlight. Some of these, probably the most ancient, convert organic material (that is, short carbon chains such as sugar) derived from decaying life into ATP and thence their own substance by the process of glycolysis (see page 70). Others can use simple *inorganic* molecules from which to generate energetic electrons and ATP. They then use the electrons, hydrogen ions, and ATP to convert CO_2, much like the final steps of photosynthesis. An intriguing account of the habits of such organisms, many of which play important roles in the ecology of our planet, may be found in *The Outer Reaches of Life* by John Postgate, Cambridge University Press, 1994.

p. 76 "The possibility that we could arise..." The 747 analogy comes from Fred Hoyle, quoted in *Origins: A Skeptic's Guide to the Creation of Life on Earth* by R. Shapiro (Summit Books, 1986).

Chapter 3. Information

p. 84 "Chemical Units of Information." The four letters are the basis of a digital system. One advantage of digital systems is that information doesn't degrade even after making multiple copies of itself. If you copy a compact disk, then make a copy of the copy, etc., the hundredth copy would sound as true as the original. This would not be true of a phonograph record or a cassette tape.

p. 97 "Copying Genes Into Messengers." In the late 1970s, scientists made the astonishing discovery that, unlike the simple state of affairs we have described that is characteristic of bacteria, genes in the larger, nucleated cells of higher organisms (eucaryotic cells) are interrupted by long sequences of nucleotides that do not code for a protein or a part of a protein. The sequences that do code for protein are called exons, while the non-coding sequences are called introns. To make a complete protein, a long RNA copy is made of the stretch of DNA containing both exons and introns. Inside the nucleus, RNA-splicing enzymes cut out the introns and link the exons together to make a messenger RNA, which leaves the nucleus and is translated into protein on ribosomes in the cytoplasm.

It is now widely hypothesized that this split nature of genes is the most ancient and that modern bacteria, to grow more efficiently, lost their introns after their exons fully developed. It appears that, out of long, meaningless stretches of DNA, certain segments (exons) evolved that coded for useful *parts* of proteins (a special shape, a special affinity, etc.).

Mechanisms then arose (RNA splicing) for uniting useful parts into final functional proteins. The advantage of such a modular system is that parts can be combined in different ways to create a great variety of proteins with diverse functions.

CHAPTER 4. MACHINERY

p. 112 "Each adaptor recognizes a particular three-letter code." Earlier (pages 84, 85) we compared nucleotides with letters and genes with paragraphs. Now we can extend the metaphor by comparing each three-letter code with a word. One word translates into one amino acid.

p. 114 "From DNA to Protein — A Multistep Process." The story of the discovery, in the 1950s, of amino activation and of transfer RNA by Mahlon Hoagland, Paul Zamecnik, and their colleagues at The Massachusetts General Hospital is told in *Toward the Habit of Truth: A Life in Science* by Mahlon Hoagland (W. W. Norton, 1990).

p. 116 "Assembling the Protein Chain." For clarity, we show one gene translated into one protein. In reality, many proteins, in their final working form, are multimers: two or more separate proteins that fit snugly together. Each protein is made from a messenger from a different gene, and these genes may not be contiguous in the DNA. But after ribosomes complete the assembly of each, they are brought together to form the final working unit.

CHAPTER 5. FEEDBACK

p. 125 "Signaling, Sensing, and Reacting." A general explanation of feedback principles by a pioneer in the field, Norbert Wiener, may be found in *The Human Use of Human Beings*, Avon Books, 1967.

"Feedback is a central feature of life." Because a feedback system treats its own state as information, some see it as similar to a mental process operating in all organisms and ecosystems — as well as in the minds of humans. Gregory Bateson advances this point of view in *Mind and Nature*.

p. 130 We owe the discovery of allostery in biological systems to the great French microbiologist and Nobel laureate Jacques Monod and his colleague Jean-Pierre Changeux. Monod's book *Chance and Necessity* (Alfred A. Knopf, 1971) is rich in the scientific and philosophical background of many of the subjects discussed in this book.

p. 134 "Controlling the Machinery that Makes the Machinery." Jacques Monod and Francois Jacob of the Pasteur Institute in Paris pioneered the exploration of regulating protein synthesis by repressors. They shared a Nobel Prize for this work in 1965.

p. 144 "Ecology Loops." The cybernetic properties of the freshwater cycle are described in Barry Commoner's *The Closing Circle*, Bantam Books, 1971.

"...ecosystems operate not as single loops but as networks..." The scientist-inventor James Lovelock has offered numerous scenarios for the earth ecosystem operating as a single, large-scale network of feedback loops (*The Ages of Gaia*, W. W. Norton, 1990).

CHAPTER 6. COMMUNITY

p. 150 "Super-organisms?" Sources of this discussion include *Three Scientists and Their Gods* by Robert Wright (Times Books, 1988) and *The Insect Societies* by E. O. Wilson (Belknap Press, 1971).

p. 152-153 "Two-faced and Slimy." Much of this information comes from *Cells and Society* by John Tyler Bonner, Princeton University Press, 1955.

p. 172 "A Chain of Command." Not all signaling originates within the embryo. In mammals, signals from the mother pass to the embryo by way of the placental bloodstream. In certain species, nurse cells adjacent to the egg signal the egg to begin its development. In honey bee populations, the queen determines the sex of bees by deciding which eggs to fertilize. Unfertilized eggs will become male; fertilized, female. In some cases, temperature acts as a signal: warm alligator eggs develop into males, cool ones into females.

CHAPTER 7. EVOLUTION

p. 178 "An Ancient Earth." In emphasizing the gradualness of geologic change, Hutton and his followers may have underestimated the role of past catastrophes in the evolution and extinction of life. Fossil records tell the story of several massive extinctions of species brought about by climate changes, impacts of asteroids, etc. After each one a great explosion of new life forms occurred, owing to the new possibilities opened up. In *Wonderful Life: The Burgess Shale and the Nature of History* (W. W. Norton, 1989), Stephen J. Gould gives a dramatic account of the explosive emergence of new life forms in the Cambrian period.

CHAPTER 7. EVOLUTION (CONTINUED)

p. 179 "Chance and selection are fundamental to any creative act." Often evolutionists find themselves in disagreement with those who believe that life is so complex and beautiful that it must have been designed. To many, the word "designed" means something like "planned in advance" — a misleading definition. Experienced designers, artists, and scientists know that design doesn't work that way. Creativity involves taking advantage of accidents, chance encounters, surprises, etc. In other words, design must access the random. If it did not, nothing new would happen.

p. 182 "Self-replicating chains." The replicator idea comes from Richard Dawkins in *The Selfish Gene*, Oxford University Press, 1989. This book and his later books *The Blind Watchmaker* (W. W. Norton, 1986) and *River Out of Eden* (Basic Books, 1995) are especially fine statements of evolutionary evidence and theory.

p. 188 "An Elephant-sized Mouse." This imaginary experiment comes from Ledyard Stebbins in *Darwin to DNA, Molecules to Humanity*, W. H. Freeman, 1982.

p. 190 "Monkeys and Word Processors." This is a good place to state that metaphor can be stretched too far, or taken too literally. The key points we use the metaphor to illustrate are (1) that *chance* pecking produces, in small steps, useful or meaningful sequences of letters — i.e., information; (2) that the information *accumulates* by a selection process; and (3) that each level of complexity can set the stage for a higher level of complexity (a process that engineers call "bootstrapping"). Clearly, the metaphor breaks down when we recognize that monkeys are using computers, a human invention; and that we humans are dictating the goal, sonnets.

p. 192 "...even the slightest advantage will take hold..." Richard Dawkins in *The Blind Watchmaker* (W. W. Norton, 1986) advances this idea most forcefully.

p. 197 "The best information sets..." Labeling a gene "better" or "worse" usually depends on how well the protein it codes for works in its particular setting. It may not be equally adaptive in another, and no environment is static. Volcanoes, earthquakes, glaciers, asteroids, and continental drift can destroy the best-adapted creatures.

p. 203 "An enzyme in the flower's cells makes pigment molecules." Assigning the flower's color to a single gene may be an oversimplification, but should convey the idea. Likewise, the white flower is meant to convey an absence of pigmentation, not a typical white flower, which may be reflecting ultraviolet light invisible to us but not to the bee.

p. 204 "Geographic Isolation." The creative possibilities inherent in small geographically isolated groups are analogous to those of such circles of artists as the French Impressionists in the late 1800s and the American Abstract Expressionists in the 1950s. Both groups were small, isolated from the mainstream museums and critics, and freely exchanging ideas within themselves. And both groups created major new movements (i.e., large changes) in a very short period of time.

p. 210 "...toward greater variety and complexity." The idea that life inevitably evolves toward greater variety and complexity is a matter of some controversy among biologists. Fossil evidence suggests that a number of organisms have remained virtually unchanged for hundreds of millions of years (for example, the horseshoe crab). Nevertheless, a general drift toward the more complex, the more varied, and the more interactive seems undeniable.

We can also see a trend toward the more abstract. In a system that builds on top of its simpler foundations, as evolution does, the "add-ons" tend to operate at a more *abstract* level of logic — i.e., more indirectly. Regulatory genes (See *Genes as Switches*, page 164) offer a good example. These genes operate on other genes — the ones that produce the worker proteins — and must have evolved after them. Further evolution then produces a regulator that controls a whole set of regulators. Such a hierarchy of control seems to be a central feature of intelligence, and it is by such layering that progressive complexity evolves.

p. 211 "High-speed evolution." The story of Peter and Rosemary Grant's exhaustive study of finches is told in the *Beak of the Finch, A Story of Evolution in Our Time*, Alfred A. Knopf, 1994.

p. 212 "A key part of the environment for organisms is other organisms." It could equally be said that the main environment for an individual gene is other genes. Indeed, it is at this molecular level that we see the most fundamental cooperation.

p. 219 "Tree of Relatedness." This phylogenetic tree is based on analyses of amino acid sequences in cytochrome c, a protein found in all the organisms shown. The study was done by Walter M. Fitch and Emanuel Margoliask and originally published in *Science* 155, 279–284, 1967. A modified version of the tree appeared in *The Mechanism of Evolution* by Francisco Ayala, *Scientific American* 239, September 1978. Our version derives from the latter. As stated by Ayala, "The numbers on the branches are the minimum number of nucleotide substitutions in the DNA of the genes [of cytochrome c] that could have given rise to observed differences in amino acid sequences."

p. 220 "Evolution of Intelligence." Excellent sources on the evolution of the brain are E. O. Wilson, *On Human Nature*, Harvard University Press, 1978, and *The Origin of Consciousness in the Breakdown of the Bicameral Mind*, Houghton Mifflin, 1976.

The idea that our brain evolved in three distinct stages was proposed by Paul D. Maclean in *Astride the Two Cultures*, edited by Harold Harris, Hutchinson, 1976. Our illustrations somewhat oversimplify in suggesting such marked divisions between reptiles, mammals, and humans. Contemporary species exhibit more subtle gradations and overlaps.

p. 222 "Cultural Evolution." For a good discussion, see Daniel C. Dennett, *Consciousness Explained*, Little, Brown, 1991.

INDEX